Numerical

Numerical Analysis

A First Year Course

by R. F. CHURCHHOUSE

Professor of Computing Mathematics
University College Cardiff

University College Cardiff Press

Copyright © 1978 University College Cardiff Press

First published 1978 in Great Britain by
University College Cardiff Press, P.O. Box 78,
Cardiff, CF1 1XL, Wales, in association with
Christopher Davies (Publishers) Ltd.,
4–5 Thomas Row, Swansea, Wales.

Printed in Wales by Salesbury Press Ltd.

All rights reserved. No part of this publication
may be reproduced, stored in a retrieval system
or transmitted, in any form or by any means,
electronic, mechanical, photocopying, recording
or otherwise, without the prior permission of
the Publishers.

ISBN 9014 2684 9

Contents

		PAGE
	PREFACE	vii
Chapter 1.	INTRODUCTION	1
Chapter 2.	DECIMAL REPRESENTATION OF NUMBERS. ROUNDING	5
Chapter 3.	ERROR AND RELATIVE ERROR	9
Chapter 4.	INTERPOLATION	17
Chapter 5.	DETECTION OF ERRORS	31
Chapter 6.	NUMERICAL INTEGRATION	39
Chapter 7.	THE SOLUTION OF NON-LINEAR EQUATIONS	53
	SOLUTIONS TO EXERCISES	71
	INDEX	73

Preface

This book is based upon the course of approximately 18 lectures on Numerical Analysis which has been given to students taking Computing at Part 1 at University College, Cardiff. It is assumed that students taking the course have an A-Level or equivalent in Mathematics and are familiar with Taylor's Theorem.

The other lectures in Part 1 Computing are devoted to programming and these are covered in a volume on Algol W Programming by Dr. W. F. B. Jones.

I am grateful to University College, Cardiff for permission to use questions from past examination papers as examples.

R. F. Churchhouse
Department of Computing Mathematics
University College, Cardiff

Chapter 1

Introduction

At school and at University students are taught how to solve certain kinds of mathematical problems by what might be loosely called "analytical methods" which include all forms of algebraic manipulation, differentiation, integration, expansion in power series, summation of infinite series etc. The armoury is vast and contains many powerful weapons which have, over the centuries, been applied with amazing ingenuity and elegance by great mathematicians such as Euler[*], Gauss[†], Jacobi[‡] and Ramanujan[§] in their solutions of problems of great difficulty. The student soon learns however that there are many problems for which no analytical solution can be found. For example he will have learned at school that there is a formula which can be used to solve any quadratic equation

$$ax^2 + bx + c = 0$$

He may learn later that there is a corresponding formula for solving the general cubic equation

$$ax^3 + bx^2 + cx + d = 0$$

and that there is also an analytical method for solving the general equation of the 4th degree. For centuries mathematicians tried to find a formula for solving the general equation of the 5th degree

$$ax^5 + bx^4 + cx^3 + dx^2 + ex + f = 0$$

but it was proved by Abel[‖] in 1823 that no such formula can

---

[*] L. Euler (1707–1783; Swiss)
[†] C. F. Gauss (1777–1855; German)
[‡] C. G. J. Jacobi (1804–1851; German)
[§] S. Ramanujan (1887–1920; Indian)
[‖] N. H. Abel (1802–1829; Norwegian)

Numerical Analysis

exist. Since equations such as this, and indeed of much higher degree, occur, how are we to solve them?

The student will also have learned how to evaluate certain classes of integrals such as

$$\int_0^t e^{-x} \, dx = 1 - e^{-t}$$

and, for $-1 \leq t \leq 1$,

$$\int_0^t \frac{dx}{\sqrt{1-x^2}} = \sin^{-1}(t)$$

the first of which follows from the simple properties of e^x and the second by means of the substitution $x = \sin \theta$. He might then try to evaluate the apparently related integral

$$\int_0^t \frac{e^{-x}}{\sqrt{1-x^2}} \, dx \quad (-1 \leq t \leq 1)$$

but no elementary solution exists. How then are we to find the value of such an integral?

Many more such examples could be given, but the point should already be clear: that there are problems for which a solution by analytical means is either impossible, or very difficult. Even when an analytical solution can be found it may be so complicated that it is quite unsuitable for computing a numerical value.

In order to solve such problems it has been found necessary to create the subject of numerical analysis, which is concerned firstly with the development and use of methods for solving mathematical problems based upon the operations of arithmetic and secondly with the study of the convergence and accuracy, and other related properties, of such methods. Some of these numerical methods were developed a long time ago by Newton[*] and others but the introduction of the computer led to a renewed interest in the subject and many new methods have been developed since 1950 which are particularly suitable for use on computers. Computers are fast, probably at least a million times faster at doing arithmetic than most human beings; they don't get bored with repetitive calculations, and they very rarely make mistakes. On

---

[*] Isaac Newton (1642–1727; English)

Introduction

the other hand they don't realize, as a human does, when a calculation is going seriously wrong so the programs need to have special checks built in to detect errors, divergence, instability, looping and so on.

Numerical analysis is a subject where practice is essential to a thorough understanding. For this reason the student is urged to make sure that he understands the worked examples in the text and tests his understanding by trying the exercises at the end of each Chapter. Most of these exercises can be done easily enough if a desk machine or pocket calculator is available; a few involve the use of a computer but even here no great programming skill is required.

Note

Details of the life and works of the mathematicians mentioned in this book can be found in books dealing with the History of Mathematics, for example: *Men of Mathematics* by E. T. Bell (Pelican Books, 2 vols. 1953).

Chapter 2

Decimal Representation of Numbers. Rounding

We shall normally represent numbers in decimal form since this is what we are accustomed to do since childhood and no alternative form offers any overwhelming advantages. Almost all computers handle numbers in binary form but this need not concern us at present and, in any case, most of the theoretical analysis is easily changed to deal with binary (or any other base, such as octal) rather than decimal.

When we represent a number such as π numerically we inevitably have to take an approximate value. The best-known fractional approximation is 22/7 but 355/113 is much better. In numerical work however we usually make use of decimal rather than fractional approximations, mainly because we can much more easily see how accurate an approximation we are using. It is known that the value of π is

$$\pi = 3 \cdot 14159265 \ldots$$

and that the digits neither terminate nor repeat (i.e. that there is no fraction that is exactly equal to π)*. It follows that if we use π in a calculation we must decide how accurate an approximation we need. Since the value of π is now known, thanks to computers, to 100,000 places† we have plenty of scope for choice! This requirement leads us to the question of what we mean by accuracy of an approximation.

---

* In fact a good deal more is known. F. Lindemann proved in 1882 that π is *transcendental*, i.e. that it is not the root of a polynomial, with integer coefficients, of *any* degree.

† "Calculation of π to 100,000 decimals", D. Shanks, J. W. Wrench, *Mathematics of Computation*, **16** (1962), 76.

Numerical Analysis

Definition 2.1

A number y is said to be an approximation which is accurate to k places of decimals to another number x if

$$|x-y| \leq \tfrac{1}{2} \times 10^{-k} \qquad (2.1)$$

Example 2.1 How accurate an approximation to π is 22/7?

Solution

$$\pi = 3\cdot 14159265\ldots$$
$$\tfrac{22}{7} = 3\cdot 14285714\ldots$$

Hence $\quad \pi - \tfrac{22}{7} = -0\cdot 0012\ldots$

and since $\quad \tfrac{1}{2}\times 10^{-2} > 0\cdot 0012 > \tfrac{1}{2}\times 10^{-3}$

we see that the approximation is correct to only 2 places of decimals.

We shall frequently abbreviate the phrase "places of decimals" to "d.p.". We now introduce the concept of *rounding*.

Definition 2.2

A number is said to be *rounded to k places of decimals* when it contains k digits after the decimal point and the absolute value of the difference between the true value and the approximate value does not exceed $\tfrac{1}{2}\times 10^{-k}$

Example 2.2. If we round π to 2 places of decimals we get $3\cdot 14$; but if we round it to 3 places of decimals we get $3\cdot 142$ (*not* $3\cdot 141$).

The definition of rounding which has just been given produces a unique result for any given number and any given value of k except when

$$|(\text{True Value}) - (\text{Approximate Value})| = \tfrac{1}{2}\times 10^{-k}$$

for in this case there will be *two* approximate values which satisfy the definition 2.2. If T denotes the true value and A the rounded value and

$$T - A = \tfrac{1}{2}\times 10^{-k}$$

then $(A + 10^{-k})$ can also be considered to be the rounded value

Decimal Representation of Numbers. Rounding

of T for
$$T-(A+10^{-k}) = -\tfrac{1}{2}\times 10^{-k}$$
In order to overcome such ambiguity some convention must be adopted. The simplest is that of always choosing the larger of the two approximate values in such circumstances. This is known as "rounding up". This has the disadvantage however that it could produce a bias in a calculation if it occurred sufficiently frequently. To avoid this various refinements have been suggested, such as alternately choosing the larger and smaller approximations, i.e. alternately "rounding up" and "rounding down". In an exceptional case this refinement could produce an even worse bias, although in practice it is unlikely to do so. A more sophisticated method is to "round up" or "round down" according to whether a pseudo-random integer generated in the computer is odd or even. This will eliminate any bias but suffers from the disadvantage that it may not be possible to repeat the calculation on the computer and obtain the same result as before.

In a computer, where the representation of numbers is in binary the rounding will of course be carried out to a fixed number of binary places, dependent on the word-size of the machine. Rounding of numbers is carried out automatically by circuitry provided by the designers and engineers who built the machine.

One of the consequences of rounding all values to a fixed number of places at every stage of a calculation is that two different answers may be obtained by performing the calculation in two different ways, as is illustrated by the following example.

Example 2.3. Form the product $(3\cdot 90+1\cdot 22)(5\cdot 46+3\cdot 59)$ (i) directly and (ii) as the sum of the four products, rounding all the numbers to 2 d.p. at each stage.

Solution (i) $(3\cdot 90+1\cdot 22)(5\cdot 46+3\cdot 59)$
$= (5\cdot 12)(9\cdot 05) = 46\cdot 34$ to 2 d.p.

(ii) $3\cdot 90\times 5\cdot 46 + 3\cdot 90\times 3\cdot 59 + 1\cdot 22\times 5\cdot 46 + 1\cdot 22\times 3\cdot 59$
$= 21\cdot 29 \quad\quad +14\cdot 00 \quad\quad +6\cdot 66 \quad\quad +4\cdot 38$
$= 46.33$

Had we performed all the intermediate calculations to 4 d.p. we

Numerical Analysis

would have obtained 46·3360 which, on rounding to 2 d.p. agrees with the first answer.

This phenomenon is an inevitable consequence of rounding. With experience numerical analysts learn how to arrange their calculations so that the effect of rounding errors is minimized.

Problems on Chapter 2

(1) The fractions $\frac{3}{2}, \frac{7}{5}, \frac{17}{12}, \frac{41}{29}$ and $\frac{99}{70}$ are all approximations to $\sqrt{2}$. If the value of $\sqrt{2}$ to 8 d.p. is 1·41421356 to how many d.p. are these five approximations accurate?

(2) How accurate an approximation to π is 355/113?

(3) Can you suggest a reason why the first method in Example 2.3 was likely to produce a more accurate result than the second method?

Chapter 3

Error and Relative Error

The accuracy of a numerical solution may be affected by errors of various kinds. We begin by describing four of the most common.

(i) *Empirical errors*

These arise when experimental data is used in the solution of the problem and the data can only be obtained with limited accuracy. Such errors may or may not be serious but in any case the accuracy with which the experimental data is known might well determine the accuracy which we should try to achieve in the solution. There is no point in trying to find a solution which is accurate to 5 d.p. when one of the experimental parameters can only be guaranteed to 2 d.p. Some examples which illustrate the effect of empirical errors on the solution of some simple problems are given below (Problems 3.1, 3.2).

(ii) *Human errors*

In hand work or desk machine work human errors are all too common. The most frequent types are:
(a) transposition of 2 digits, e.g. writing 17236 instead of 17326;
(b) repetition of the wrong digit, e.g. writing 39967 instead of 39667.
(c) forgetting a negative sign in front of a number, particularly when working with tables.

When using computers such errors will not occur in the machine itself but they might occur in the preparation of the input data. Some numerical methods for the solution of problems contain built-in checks to guard against errors of this kind. Such checks should always be incorporated whenever possible, for example in

Numerical Analysis

the solution by hand of sets of linear equations using Gaussian elimination.

(iii) *Rounding errors*

In all numerical work whether carried out by hand, on a desk machine, or on a computer we will have to work with a limited number of decimal places. Consequently we will be continually rounding all the numbers which occur during the course of the calculation. We saw a simple example of the effect of this in Example 2.3. Since some of the numbers will be rounded up and others will be rounded down the accumulated rounding error at the end of a long calculation may be positive or negative (it is unlikely to be zero!) and we may have only the vaguest idea of its size. Suppose for example that we are adding up n numbers and that each number is rounded to k d.p. Then the accumulated rounding error obviously satisfies

$$-\frac{n}{2} \times 10^{-k} \leq \text{accum. rounding error} \leq \frac{n}{2} \times 10^{-k}.$$

If n is at all large it is extremely unlikely that the accumulated rounding error will be anywhere near $(n/2) \times 10^{-k}$ in absolute value. For example if we evaluate the sum

$$\sum_{m=11}^{20} \frac{1}{m}$$

working at every stage to 3 d.p. we obtain the value 0·670. The maximum accumulated rounding error is $5 \times 10^{-3} = 0·005$ so we deduce that

$$0·665 \leq \sum_{m=11}^{20} \frac{1}{m} \leq 0·675$$

In fact the sum $= 0·66877$ to 5 d.p. so the accumulated rounding error is less than 0·00125, i.e. less than one quarter of its theoretical maximum. This is much more in line with what we should expect on the basis of a theorem in the theory of probability which tells us that the accumulated rounding error in the case of n additions will increase in proportion to $\sqrt{n}$ (not n).

When the calculation involves multiplications or divisions the prediction of the approximate size of the accumulated rounding error is often a problem of considerable difficulty.

Some further evidence relating to the problem of accumulated rounding errors will be found in the Problems at the end of this Chapter.

(iv) *Truncation errors*

We frequently have to evaluate the sum of an infinite series or find a value for an integral over a finite or infinite range. In such cases we will necessarily introduce a truncation error for
(a) unless there is a formula for the sum of the series we will have to be content with summing a *finite* number of terms, i.e. we replace

$$\sum_{n=1}^{\infty} a_n \quad \text{by} \quad \sum_{n=1}^{N} a_n$$

The truncation error in this case is the sum of the terms which we have neglected, viz.

$$\text{Truncation Error} = \sum_{n=N+1}^{\infty} a_n$$

(b) In the case of a finite integral which we cannot evaluate exactly by analytical means we must use a formula for numerical integration. This formula will only give an approximate value and there will be an error term which will probably depend upon some power of the "step-size" (h) and the derivative of some order of the function being integrated. We shall consider such matters in detail in Chapter 6.
(c) If the range of integration is infinite we may have to replace it by a finite range, in which case we will have a truncation error composed of two parts, one corresponding to (a) above and the other corresponding to (b). In certain cases such as $\int_0^{\infty} e^{-x} f(x) \, dx$ integration formulae have been specially developed for integration over the infinite range and the truncation error is then simply one of type (b).

3.1. Absolute Error and Relative Error

The *Error*, or as it is also called the *Absolute Error*, in an approximation to a number is defined as follows:

Definition

$$\text{Error} = (\text{True Value}) - (\text{Approximate Value})$$

Sometimes a more meaningful measure of the accuracy of an

Numerical Analysis

approximation is given by the *Relative Error*.

Definition

$$\text{Relative Error} = \frac{\text{Error}}{\text{True Value}}$$

The Relative Error often provides a much better guide to the seriousness of an error than the Absolute Error provides. This is particularly so when the True Value is either very small or very large. Thus an error of 1000 may look serious but if the True Value is the distance of the Earth from the Sun in miles (about 93×10^6) the Relative Error is only about 10^{-5}.

Example 3.1. Two parameters x, y have been estimated experimentally, $x = 2 + \varepsilon_1$, $y = 3 + \varepsilon_2$ where $|\varepsilon_1| < 0 \cdot 1$, $|\varepsilon_2| < 0 \cdot 2$. Find bounds on the value of the product xy.

Solution
(i) we first obtain an upper bound:

$$xy = (2 + \varepsilon_1)(3 + \varepsilon_2)$$
$$\leqslant (2 + |\varepsilon_1|)(3 + |\varepsilon_2|)$$
$$< (2 \cdot 1)(3 \cdot 2) \qquad = 6 \cdot 72$$

(ii) for the lower bound:

$$xy \geqslant (2 - |\varepsilon_1|)(3 - |\varepsilon_2|)$$
$$> (1 \cdot 9)(2 \cdot 8) \qquad = 5 \cdot 32$$

Hence all we can guarantee is that

$$5 \cdot 32 < xy < 6 \cdot 72$$

Example 3.2. Under the same conditions as in Example 3.1 what bounds can be set upon the value of y/x?

Solution (i)

$$\frac{y}{x} = \frac{3 + \varepsilon_2}{2 + \varepsilon_1}$$
$$\leqslant \frac{3 + |\varepsilon_2|}{2 - |\varepsilon_1|}$$
$$< \frac{3 \cdot 2}{1 \cdot 9} < 1 \cdot 69 \text{ (to 2 d.p.)}$$

(ii)
$$\frac{y}{x} \geq \frac{3-|\varepsilon_2|}{2+|\varepsilon_1|}$$
$$> \frac{2 \cdot 8}{2 \cdot 1} > 1 \cdot 33 \text{ (to 2 d.p.)}$$

So, all we can say is that
$$1 \cdot 33 < y/x < 1 \cdot 69.$$

Example 3.3. (*Truncation Error*) If we wish to produce a table of values of e^x for $0 \leq x \leq \frac{1}{2}$ using the Taylor Series

$$e^x = \sum_{n=0}^{\infty} \frac{x^n}{n!}$$

how many terms must we use in order to be sure that our results are correct to 4 d.p.?

Solution. Suppose we take N terms of the Taylor Series, i.e. we take
$$e^x = \sum_{n=0}^{N-1} \frac{x^n}{n!}$$

then the Truncation Error (T.E.) is

$$\sum_{n=N}^{\infty} \frac{x^n}{n!}$$

i.e.
$$\text{T.E.} = \frac{x^N}{N!}\left(1 + \frac{x}{(N+1)} + \frac{x^2}{(N+1)(N+2)} + \cdots\right)$$

We need to estimate how big T.E. can be and since we wish to guarantee that our results are correct to 4 d.p. we may, if necessary *overestimate* T.E., but we must not *underestimate* it.

Now, since $N \geq 0$

$$1 + \frac{x}{(N+1)} + \frac{x^2}{(N+1)(N+2)} + \cdots \leq 1 + \frac{x}{1} + \frac{x^2}{1 \cdot 2} + \frac{x^3}{1 \cdot 2 \cdot 3} + \cdots = e^x$$

and so, because $0 \leq x \leq \frac{1}{2}$

$$\text{T.E.} \leq \frac{(\frac{1}{2})^N}{N!} \cdot e^{\frac{1}{2}} < \frac{2(\frac{1}{2})^N}{N!}$$

Numerical Analysis

For accuracy to 4 d.p. we require

$$|\text{T.E.}| \leq \tfrac{1}{2} \times 10^{-4} = \frac{1}{20{,}000}$$

and so we can guarantee this if N is chosen so that

$$\frac{2(\tfrac{1}{2})^N}{N!} < \frac{1}{20{,}000}$$

We now try some values for N:
$N = 5$:

$$\frac{2(\tfrac{1}{2})^5}{5!} = \frac{1}{1920} > \frac{1}{20{,}000}$$

$N = 6$:

$$\frac{2(\tfrac{1}{2})^6}{6!} = \frac{1}{23{,}040} < \frac{1}{20{,}000}$$

Thus 6 terms will suffice.

3.2. Estimation of Truncation Error

Example 3.3 illustrates how we tackle the problem of estimating truncation error when we are faced with a series which we cannot sum analytically. In such cases we must replace the terms of the series by bigger terms of another series which we can sum. Thus if the T.E. is $\sum_{k=N}^{\infty} a_k$ and if $|a_n| < b_n$ and we can sum $\sum_{k=N}^{\infty} b_k$ then we use the estimate

$$\text{T.E.} = \sum_{k=N}^{\infty} a_k < \sum_{k=N}^{\infty} b_k$$

Similarly we can deal with the truncation error involving an integral, i.e. we write

$$\text{T.E.} = \int_N^{\infty} f(x)\, dx < \int_N^{\infty} g(x)\, dx$$

where $|f(x)| < g(x)$, and we can evaluate $\int_N^{\infty} g(x)\, dx$ directly.

Problems on Chapter 3

(1) The 10 variables $x_1, \ldots, x_{10}$ have been measured experimentally and it has been found that $x_k = k$ with a relative error of 10%. What bounds can be set upon the values of

   (i) $\sum_{i=1}^{10} x_i^2$,   (ii) $\prod_{i=1}^{10} x_i$,   (iii) $\sum_{k=1}^{10} kx_k^{-1}$?

(2) What bounds can be set on the roots of the quadratic equation

$$x^2 - ax + b = 0$$

when $a = 5 + \varepsilon_1$, $b = 6 + \varepsilon_2$ and $|\varepsilon_1| < 0.05$, $|\varepsilon_2| < 0.02$?

(3) Find the value of $\sum_{n=1}^{20} \sqrt{n}$ when all numbers are rounded at each stage to

   (i) 0 places of decimals,   (ii) 1 place,   (iii) 2 places.

(4) Estimate the number of terms of the series $\sum_{n=0}^{\infty} (\tfrac{1}{2})^{n^2}$ that we must take if we are to find its value correct to (i) 3 d.p, (ii) 10 d.p., (iii) 1000 d.p.

(5) Given that the value of $\sqrt{3}$ is known approximately compare the errors in computing (i) $1/(2+\sqrt{3})^4$ and (ii) the equivalent expression $97 - 56\sqrt{3}$. (Hint: replace $\sqrt{3}$ in both expressions by $\sqrt{3} + h$, where h is the error, and ignore h^2, h^3 etc.)

Chapter 4

Interpolation

Many of the elementary problems we were given at school involved the use of logarithms or the trigonometric functions, sine, cosine, tangent etc. Tables of such functions are readily available, typically to 4 d.p. Suppose however that we have access only to a table of logs of the integers from 1 to 100, how do we estimate the value of, say, log 87·62? An obvious method is to use the value

$$\log 87 + \frac{62}{100}(\log 88 - \log 87)$$

as an approximation. If we do this we obtain the value 1·9426, which is correct to 4 d.p.

It is instructive to interpret this method graphically. Consider the curve $y = \log x$ and let P, Q be the points (87, log 87) and (88, log 88) respectively. Then P, Q lie on the curve. Let R be the point (87·62, log 87·62) then R also lies on the curve and our problem is to find a point as close to R as possible given only the positions of P, Q. The method suggested above is equivalent to drawing the straight line joining P to Q and taking R to be the point where this line crosses the vertical line $x = 87·62$.

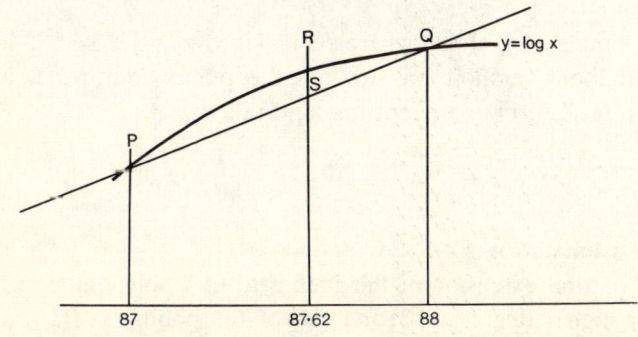

17

Numerical Analysis

The point so found, S in the diagram, will not coincide with R but will be close to it if P and Q are not too far apart and if the curve joining them is almost a straight line. These remarks contain several vague phrases but although they can be put into a rigorous mathematical form it is not necessary for us to do so in the present context. We might however see how our result would be affected if we had taken P to be the point $(80, \log 80)$ and Q to be $(90, \log 90)$ then our estimate for $\log 87 \cdot 62$ would be

$$\log 80 + 0 \cdot 762 (\log 90 - \log 80)$$

which gives the value $1 \cdot 9421$ to 4 d.p. This result is not even correct to 3 d.p. The reason for this marked decline in accuracy is intuitively obvious: we have based the calculation on two points P, Q which are in a sense, "too far apart" for the chord PQ to be a close approximation to the curve of $\log x$ between P and Q.

The problem of estimating the value of a function at a point R given the values of that function at a set of points on both sides of R is one of the most fundamental in numerical analysis and is known as the problem of *interpolation*. If, as sometimes happens, the values of the function are known only at a set of points *on one side* of R we are faced with the intrinsically more difficult problem of *extrapolation*.

In order to solve the problem of interpolation an approach such as the following seems almost obvious: suppose that the values of a function, $f(x)$ are known only at a set of points $x_1, x_2, \ldots, x_n$. We wish to estimate the value of the function at some point, y, which is not one of the x_i, but which lies somewhere between them so that, say,

$$x_1 < x_2 < \ldots < x_k < y < x_{k+1} < \ldots < x_n.$$

In the simple case above where $f(x) = \log x$, $x_1 = 87$, $x_2 = 88$ and $y = 87 \cdot 62$ our method was to join the points $P(x_k, f(x_k))$ and $Q(x_{k+1}, f(x_{k+1}))$ by a straight line and take

$$f(y) = f(x_k) + \frac{y - x_k}{x_{k+1} - x_k} (f(x_{k+1}) - f(x_k))$$

as our estimate for $f(y)$.

The natural extension of this idea is to fit a polynomial of the 2nd or higher degree to 3 or more of the points $(x_i, f(x_i))$ and then use this polynomial to estimate the value of $f(y)$. We are therefore led to the problem of fitting a polynomial to a set of given points and we now see how this can be done.

4.1. Fitting the Interpolation Polynomial

The most obvious method for fitting a polynomial is illustrated in the following simple example.

Example 4.1. Find a polynomial which fits the following data

$$x = 0 \quad 1 \quad 5$$
$$f(x) = 6 \quad 2 \quad 6$$

Solution. Since we have three points we can fit a polynomial containing three coefficients, i.e. a polynomial of degree two. Let this polynomial be

$$f(x) = ax^2 + bx + c$$

substituting the given data we obtain three linear equations in the three unknowns a, b, c, viz.

$$c = 6$$
$$a + b + c = 2$$
$$25a + 5b + c = 6$$

The solution of these equations is $a = 1$, $b = -5$, $c = 6$ so that the required polynomial is $x^2 - 5x + 6$.

Similarly we can fit a polynomial of degree at most m to the values $f(x_i)$ of a function at $m+1$ points, x_i $(i = 1, 2, \ldots, m+1)$ by solving the system of $(m+1)$ linear equations

$$\sum_{r=0}^{m} a_r x_i^r = f(x_i) \quad (i = 1, 2, \ldots, m+1)$$

in the $(m+1)$ unknowns $a_0, a_1, \ldots, a_m$ the required polynomial being

$$\sum_{r=0}^{m} a_r x^r.$$

Numerical Analysis

Whilst this is a perfectly satisfactory method for small values of m (say, $m < 4$) it rapidly becomes very tedious and time-consuming as m increases since the work involved in solving a system of m equations is proportional to m^3. If a computer is being used the method is quite acceptable for much larger values of m but by hand most people would hesitate to go even as far as $m = 10$. We shall discuss in sections 4.2 and 4.5 below two other methods for finding the interpolation polynomial. The first of these, due to Lagrange*, has the merit of simplicity; in particular it is easy to see that the polynomial which we construct really does fit the given data. The second method, due to Newton, is generally regarded as the one best suited to hand calculation; it is, however, based upon the use of divided differences and these we introduce in section 4.4.

4.2. The Lagrange Polynomial

The idea behind this method should be clear from the following simple example.

Example 4.2. Fit a polynomial to the following data

$$x = \quad 1 \quad 3 \quad 4$$
$$f(x) = -3 \quad 3 \quad 9$$

Solution. We consider the polynomial

$$f(x) = a(x-3)(x-4) + b(x-1)(x-4) + c(x-1)(x-3)$$

and observe that when we put $x = 1$ only the term involving a is non-zero, at $x = 3$ only the term involving b is non-zero and at $x = 4$ only the term involving c is non-zero. Thus we can find the values of a, b, c very easily.

Putting $x = 1$:

$$a(-2)(-3) = f(1) = -3, \quad a = -\tfrac{1}{2}$$

Putting $x = 3$:

$$b(2)(-1) = f(3) = 3, \quad b = -\tfrac{3}{2}$$

* J. L. Lagrange (1736–1813; French)

Putting $x = 4$:
$$c(3)(1) = f(4) = 9, \quad c = 3$$
Hence $f(x) = -\frac{1}{2}(x-3)(x-4) - \frac{3}{2}(x-1)(x-4) + 3(x-1)(x-3)$
We can now simplify $f(x)$ if we wish, giving
$$f(x) = x^2 - x - 3$$
but this may not be either necessary or desirable if $f(x)$ is required only for the purpose of interpolation.

We now formally introduce the Lagrange Polynomial in

THEOREM. There exists a polynomial $L(x)$ of degree at most $(k-1)$ which takes a given set of values $f(x_i)$ at a given set of points x_i $(i = 1, 2, \ldots, k)$.

Proof. Consider the polynomial
$$L(x) = \sum_{i=1}^{k} f(x_i) \prod_{\substack{j=1 \\ j \neq i}}^{k} \left(\frac{x - x_j}{x_i - x_j} \right)$$

$L(x)$ consists of the sum of k polynomials each of which is of degree at most $(k-1)$. It follows that $L(x)$ is itself a polynomial of degree at most $(k-1)$.

Let x_r be one of the k points x_i $(i = 1, 2, \ldots, k)$. Then $L(x_r)$ consists of the sum of k terms of which $(k-1)$ are zero since they contain the factor $(x_r - x_r)$; the only non-zero term is where $i = r$ in the definition above and this term is
$$f(x_r) \prod_{\substack{j=1 \\ j \neq r}}^{k} \left(\frac{x_r - x_j}{x_r - x_j} \right) = f(x_r)$$
so that $L(x_r) = f(x_r)$, i.e. $L(x)$ takes the values $f(x_i)$ at the points x_i $(i = 1, \ldots, k)$.

This completes the proof. Q.E.D.

The polynomial $L(x)$ is called the Lagrange Interpolation Polynomial or the Lagrange Collocation Polynomial.

Example 4.3. Find a polynomial of degree two which takes the same values as x^3 at $x = 0, 1, 2$ and use it to interpolate a value at $x = 1 \cdot 5$

Numerical Analysis

Solution
$$x = 0 \quad 1 \quad 2$$
$$f(x) = 0 \quad 1 \quad 8$$

The Lagrange polynomial is

$$L(x) = 0\left(\frac{x-1}{0-1}\right)\left(\frac{x-2}{0-2}\right) + 1\left(\frac{x}{1}\right)\left(\frac{x-2}{1-2}\right) + 8\left(\frac{x}{2}\right)\left(\frac{x-1}{2-1}\right)$$

i.e.
$$L(x) = x(3x - 2)$$

and so the interpolated value at $x = 1·5$ is $L(1·5) = 3·75$. The exact value in this case is of course

$$(1·5)^3 = 3·375$$

4.3. Forward Differences

A concept of fundamental importance in numerical analysis in general and particularly so in the problems of interpolation and error detection is that of the forward differences of a function at a set of points. Initially we shall suppose that the points are equally spaced that is $x_{r+1} - x_r = h$ for $r = 1, \ldots, (n-1)$.

Definition

Given the values $f(x)$ of a function at a set of equally spaced points x_i $(i = 1, 2, \ldots, n)$ where $x_1 < x_2 < x_3 \ldots < x_n$ and $x_{i+1} - x_i = h$ we define the First Forward Difference of $f(x)$ at x_i to be $f(x_{i+1}) - f(x_i)$ and denote it by Δf_i.

From this definition we proceed to define the higher order forward differences:

Definition
For $k \geq 2$ the k-th forward difference of $f(x)$ at x_i is defined by

$$\Delta^k f_i = \Delta^{k-1} f(x_{i+1}) - \Delta^{k-1} f(x_i)$$

where
$$\Delta^1 f_i = \Delta f_i.$$

Example 4.4. Show that

$$\Delta^2 f_i = f(x_{i+2}) - 2f(x_{i+1}) + f(x_i)$$

Proof.
$$\Delta^2 f_i = \Delta f_{i+1} - \Delta f_i$$
$$= (f_{i+2} - f_{i+1}) - (f_{i+1} - f_i)$$
$$= f_{i+2} - 2f_{i+1} + f_i$$
i.e. $\quad \Delta^2 f_i = f(x_{i+2}) - 2f(x_{i+1}) + f(x_i)$ Q.E.D.

An important property of polynomials of degree k is that their k-th forward differences are constant, as the next example indicates, and as we shall prove in the theorem which follows.

Example 4.5. Tabulate the values of $f(x) = x^3$ at $x = 0(1)6$ and form the table of differences.

Note: $x = 0(1)6$ means "x starts with the value 0 and increases in steps of 1 up to, and including, 6".

Solution

$x = 0$		1		2		3		4		5		6
$f(x) = 0$		1		8		27		64		125		216
Δf	1		7		19		37		61		91	
$\Delta^2 f$		6		12		18		24		30		
$\Delta^3 f$			6		6		6		6				
$\Delta^4 f$				0		0		0					

Thus the 3rd forward difference of x^3 is constant. The value of the constant depends upon the interval at which we tabulate x^3. The 4th forward difference is identically zero for all values of the interval size.

THEOREM. If $f(x)$ is a polynomial of degree k tabulated at a set of equally spaced points x_i ($i = 1, 2, \ldots$) then $\Delta^k f_i$ is a constant and $\Delta^{k+1} f_i \equiv 0$.

Proof. Let the interval between consecutive points be h. Let x_i be one of the points and let
$$f(x) = a_k x^k + a_{k-1} x^{k-1} + \ldots + a_0$$
Then $\quad \Delta f_i = f(x_i + h) - f(x_i)$
$$= \sum_{r=0}^{k} a_r \{(x_i + h)^r - x_i^r\}$$
$$= \sum_{r=0}^{k-1} b_r x_i^r$$

where $b_0, b_1, \ldots, b_{k-1}$ involve h, but not x_i, because there are only two terms which involve x_i^k in the expression above and these cancel each other. We have therefore proved that

Δ(polynomial of degree k) = polynomial of degree $(k-1)$

Hence:

Δ^2(polynomial of degree k) = polynomial of degree $(k-2)$

and so on until

Δ^k(polynomial of degree k) = polynomial of degree 0 = constant

and finally, Δ^{k+1} (polynomial of degree k) $\equiv 0$ Q.E.D.

4.4. Divided Differences

If we have a function which is tabulated at a set of points x_i which are not equally spaced there is not much point in using forward differences since important properties of functions, such as the one just proved for polynomials, will no longer hold. Fortunately a simple generalization of the concept of forward difference not only enables us to recover these properties but also to develop a very powerful method for dealing with a wide range of problems.

Definition

If the values of a function $f(x)$ are given at two points x_1, x_2 the *first divided difference* of $f(x)$ at x_1 and x_2 is defined to be

$$\frac{f(x_1) - f(x_2)}{x_1 - x_2}$$

and is denoted by $[x_1 x_2]$.

Expressed in trigonometric terms $[x_1 x_2]$ is the tangent of the angle which the line joining the points $(x_1, f(x_1))$ and $(x_2, f(x_2))$ makes with the x-axis.

By taking the first divided difference of a pair of related divided differences we form the *second divided difference*

$$[x_1 x_2 x_3] = \frac{[x_2 x_3] - [x_1 x_2]}{x_3 - x_1}$$

Interpolation

More generally we define the k-th divided difference in terms of the $(k-1)$st for $k \geq 2$.

Definition

For $k \geq 2$ the k-th divided difference at a set of points $x_1, x_2, \ldots, x_k$ is defined by

$$[x_1 x_2 \ldots x_k] = \frac{[x_2 x_3 \ldots x_k] - [x_1 x_2 \ldots x_{k-1}]}{x_k - x_1}$$

Note: The points $x_1, x_2, \ldots, x_k$ may be selected in arbitrary order, i.e. we are not assuming that $x_1 < x_2 \ldots < x_k$ although this will frequently be the case. It is perfectly valid to use $[x_2 x_4]$ or $[x_4 x_1 x_7]$ for example.

Example 4.6. Form the divided difference table of $f(x) = x^4$ at $x = 0, 1, 3, 6, 7, 10$.

$x =$	0	1	3	6	7	10
$f(x) =$	0	1	81	1296	2401	10000
1st D.D. =		1	40	405	1105	2533
2nd D.D. =			13	73	175	357
3rd D.D. =				10	17	26
4th D.D. =					1	1
5th D.D. =						0

4.5. Newton's Interpolation Formula

THEOREM. If a function $f(x)$ takes values $f(x_i)$ at a set of points x_i ($i = 1, 2, \ldots, n$), not necessarily equally spaced, then

$$f(x) = f(x_1) + (x - x_1)[x_1 x_2] + (x - x_1)(x - x_2)[x_1 x_2 x_3] + \ldots$$
$$\ldots + (x - x_1)(x - x_2) \ldots (x - x_{n-1})[x_1 x_2 \ldots x_n]$$
$$+ (x - x_1)(x - x_2) \ldots (x - x_n)[xx_1 \ldots x_n].$$

Proof.
(i) By definition

$$[xx_1] = \frac{f(x) - f(x_1)}{x - x_1}$$

and so $f(x) = f(x_1) + (x - x_1)[xx_1]$ and this agrees with the statement of the theorem when $n = 1$. The theorem is therefore true when $n = 1$.

(ii) Suppose that the theorem has been proved for n points $x_1, x_2, \ldots, x_n$. Then since

$$[xx_1x_2 \ldots x_{n+1}] = \frac{[x_1x_2 \ldots x_{n+1}] - [xx_1x_2 \ldots x_n]}{x_{n+1} - x}$$

we have

$$[xx_1x_2 \ldots x_n] = [x_1x_2 \ldots x_{n+1}] + (x - x_{n+1})[xx_1 \ldots x_{n+1}].$$

If we now substitute for $[xx_1x_2 \ldots x_n]$ in the statement of the theorem we find that

$$\begin{aligned} f(x) = &\, f(x_1) + (x - x_1)[x_1x_2] + (x - x_1)(x - x_2)[x_1x_2x_3] \\ &\, + \ldots + (x - x_1)(x - x_2) \ldots (x - x_n)[x_1x_2 \ldots x_{n+1}] \\ &\, + (x - x_1) \ldots (x - x_{n+1})[xx_1x_2 \ldots x_{n+1}], \end{aligned}$$

and this agrees with the statement of the theorem.

Thus we have proved that if the theorem is true for n points it is also true for $(n+1)$ points and since we have seen that it is true when $n = 1$ the theorem is proved for all positive integer values of n by induction. Q.E.D.

The theorem provides the basis for a method of interpolation due to Newton. In the first place, as we show below, the higher order divided differences of a polynomial are zero so that the formula just proved can be used to determine which polynomial exactly fits a set of points. Secondly if the higher order divided differences of a tabulated function are small then the formula can be used to provide an approximation to the function and so can be used for the purpose of interpolation.

Our assertion that the k-th order divided differences of a polynomial of degree $(k-1)$ are zero is proved easily by observing that if $f(x) = x^n$ then

$$[x_1x_2] = \frac{x_1^n - x_2^n}{x_1 - x_2} = x_1^{n-1} + x_1^{n-2}x_2 + \ldots + x_2^{n-1}$$

so that the effect of taking the divided difference is to reduce the degree by 1. It follows that if we take the divided difference of

Interpolation

x^{k-1} ($k-1$) times we reduce the degree to zero, i.e. we are left with a constant and taking the divided difference once more we are left with zero.

As an example of the use of Newton's Formula in fitting a polynomial to a set of data consider again the problem given in Example 4.2.

Example 4.6. Use Newton's Method to fit a polynomial to the data

$$x = 1 \quad 3 \quad 4$$
$$f(x) = -3 \quad 3 \quad 9$$

Solution. The divided difference table is:

$$f(x) = -3 \quad\quad 3 \quad\quad 9$$
$$\text{1st D.D.} = \quad\quad 3 \quad\quad 6$$
$$\text{2nd D.D.} = \quad\quad\quad 1$$

and all higher order divided differences are taken to be zero. The theorem then gives (taking $x_1 = 1$, $x_2 = 3$, $x_3 = 4$)

$$f(x) = -3 + 3(x-1) + 1(x-1)(x-3)$$
$$= x^2 - x - 3$$

as we found before.

As a slightly more difficult example we fit a cubic:

Example 4.7. Fit a cubic to the data

$$x = 0 \quad 1 \quad 3 \quad 6$$
$$f(x) = -7 \quad -8 \quad 14 \quad 197$$

using Newton's Method.

Solution

$$x = 0 \quad\quad 1 \quad\quad 3 \quad\quad 6$$
$$f(x) = -7 \quad\quad -8 \quad\quad 14 \quad\quad 197$$
$$\text{1st D.D.} = \quad -1 \quad\quad 11 \quad\quad 61$$
$$\text{2nd D.D.} = \quad\quad 4 \quad\quad 10$$
$$\text{3rd D.D.} = \quad\quad\quad 1$$

Numerical Analysis

Taking $x_1 = 0$, $x_1 = 1$, $x_2 = 3$, $x_4 = 6$ Newton's formula then gives:

$$f(x) = -7 - 1(x) + 4(x)(x-1) + (x)(x-1)(x-3)$$

i.e.
$$f(x) = x^3 - 2x - 7$$

It is salutary to try to solve this same problem using the Lagrange Method. The advantages of Newton's Method for hand calculation will then become apparent.

4.6. Interpolation using Newton's Formula

If $f(x)$ is tabulated at a set of points x_i ($i = 1, 2, \ldots, n$) and the higher order divided differences are small we can use Newton's formula in the approximate form

$$f(x) \doteqdot f(x_1) + (x - x_1)[x_1 x_2] + (x - x_1)(x - x_2)[x_1 x_2 x_3]$$
$$+ \ldots + (x - x_1) \ldots (x - x_k)[x_1 x_2 \ldots x_{k+1}]$$

(where we neglect all divided differences of order higher than the k-th) to interpolate values in the table.

Example 4.8. From the values of $\log_{10} x$ at $x = 2, 4, 5, 8, 9, 10$ obtain an interpolated value for $\log_{10} 6$ using Newton's Formula, work to 4 d.p.

Solution

$x =$	2	4	5	8	9	10
$\log x =$	0·3010	0·6021	0·6990	0·9031	0·9542	1·0000
1st D.D. =		0·1506	0·0969	0·0680	0·0511	0·0458
2nd D.D. =			−0·0179	−0·0073	−0·0042	−0·0027
3rd D.D. =				0·0018	0·0006	0·0003
4th D.D. =					−0·0002	−0·0001

The 5th D.D. is zero to 4 d.p. Rounding errors in the 3rd and 4th D.D. might well prove to be significant since an error of 0·0001 corresponds to a relative error of up to 100%! However we apply Newton's Formula choosing our points as close to $x = 6$ as we can, viz.

$$x_1 = 5, \quad x_2 = 4, \quad x_3 = 8, \quad x_4 = 9$$

but avoid the use of the 4th D.D. as too unreliable. This gives:

$$\log_{10} 6 = \log_{10} 5 + (6-5)[5,4] + (6-5)(6-4)[5,4,8]$$
$$+ (6-5)(6-4)(6-8)[5,4,8,9]$$

i.e. $\log_{10} 6 = 0.6990 + (1)(0.0969) + (1)(2)(-0.0073)$
$$+ (1)(2)(-2)(0.0006)$$
$$0.6990 + 0.0969 - 0.0146 - 0.0024$$

i.e. $\log_{10} 6 = 0.7789$

The correct result to 4 d.p. is 0.7782.

In this case, of course, the fact that the function in the table is the logarithm enables us to deduce a value for $\log_{10} 6$ quite easily since we are given $\log_{10} 4$ and $\log_{10} 9$, hence

$$\log_{10} 6 = \tfrac{1}{2}(\log 4 + \log 9) = 0.7782 \text{ to 4 d.p.}$$

If the function in the table is known to possess some special property, such as the functional equation of the logarithm

$$\log(x) + \log(y) = \log(xy)$$

then we should make use of it whenever possible. Such opportunities will only rarely occur however. Had we wanted a value for, say, $\log_{10} 6.1$ in Example 4 no simple use of the functional equation would have produced it whereas Newton's formula leads to

$$\log_{10} 6.1 = 0.6990 + (1.1)(0.0969) + (1.1)(2.1)(-0.0073)$$
$$+ (1.1)(2.1)(-1.9)(0.0006)$$
$$= 0.6990 + 0.1066 - 0.0169 - 0.0026$$
$$= 0.7861$$

The correct result to 4 d.p. is 0.7853.

Problems on Chapter 4

(1) Fit a polynomial to the following data

$$x = 0 \quad 1 \quad 3$$
$$f(x) = 0 \quad \tfrac{1}{2} \quad 1$$

and use it to interpolate values for $f(1.5)$ and $f(2)$.

Numerical Analysis

(2) Show that the data in Problem 1 fits the function $f(x) = \sin(\pi x/6)$. Under this interpretation what are the values of $f(1\cdot 5)$ and $f(2)$ and how do they compare with the values found above?

(3) Insert the additional information that when $x = 2$, $f(x) = 0\cdot 8660$ into the data of Problem 1 and fit a cubic to the four points. Use this cubic to interpolate a value at $x = 1\cdot 5$. Compare with the known value of $\sin(\pi/4)$.

(4) Fit a polynomial to the following data

$$x = -4 \quad -2 \quad 0 \quad 1 \quad 3 \quad 5$$
$$f(x) = 180 \quad 0 \quad 4 \quad 0 \quad 40 \quad 504$$

and use the polynomial to find a value for $f(2\cdot 4)$.

(Cardiff Part 1, 1972).

Chapter 5

Detection of Errors

Finite differences can be used to detect, and occasionally to correct, errors in tables of values of functions which have sufficiently smooth behaviour. For if a function is sufficiently smooth its higher order differences should exhibit some pattern, such as slowly increasing or slowly decreasing, and any error in the table of values may destroy the pattern and this may enable us to pinpoint the value which is in error. We begin by recalling that the k-th forward difference of a polynomial of degree $(k-1)$ is zero—so if the function with which we are concerned is a polynomial its differences above a certain order will be everywhere zero: so in this case the pattern is very simple—it is a line of zeros.

If the function is not a polynomial no differences of any order will ever be zero everywhere (for if they were we could use Newton's Formula to fit an *exact* polynomial—and this contradicts the assertion that the function is not a polynomial). Let us see what sort of patterns we obtain if we form the difference table of a smooth function, such as $\sin(x)$.

Example 5.1. Form the difference table up to Δ^3 for $\sin(x)$ for $x = 0(0\cdot1)0\cdot8$

Solution

$x=$	0	0·1	0·2	0·3	0·4	0·5	0·6	0·7	0·8
$f(x)=$	0	0·0998	0·1987	0·2955	0·3894	0·4794	0·5646	0·6442	0·7173
$\Delta f=$	0·0998	0·0989	0·0968	0·0939	0·0900	0·0852	0·0796	0·0731	
$\Delta^2 f=$	−0·0009	−0·0021	−0·0029	−0·0039	−0·0048	−0·0056	−0·0065		
$\Delta^3 f=$	−0·0012	−0·0008	−0·0010	−0·0009	−0·0008	−0·0009			

We notice that all the values in Δf are positive and decreasing; that the values in $\Delta^2 f$ are negative and decreasing and the values in $\Delta^3 f$ are negative, nearly constant but slightly erratic. The erratic behaviour in the Δ^3 line is caused by rounding errors. We shall examine this aspect of a difference table later (section 5.1); for the present we simply note that it would be unwise to use $\Delta^4 f$ for any purposes in Example 5.1 since the values would be mathematically meaningless.

We now make a deliberate error in the table of values of sin x above and see what effect this has on the difference table.

Example 5.2. In the table of values above make a deliberate error by writing $\sin(0.4) = 0.3984$ and note the errors that this causes in the lines $\Delta f, \Delta^2 f, \Delta^3 f$.

Solution

$x = 0$	0·1	0·2	0·3	0·4	0·5	0·6	0·7	0·8
$f(x) = 0$	0·0998	0·1987	0·2955	0·3984	0·4794	0·5646	0·6442	0·7173
$\Delta f =$	0·0998	0·0989	0·0968	0·1029	0·0810	0·0852	0·0796	0·0731
$\Delta^2 f =$	−0·0009	−0·0021	+0·0061	−0·0219	+0·0042	−0·0056	−0·0065	
$\Delta^3 f =$		−0·0012	+0·0082	−0·0280	+0·0261	−0·0098	−0·0009	

We note that there are now two errors in the Δf row, three in the $\Delta^2 f$ row and four in the $\Delta^3 f$ row. If we draw lines as shown in the diagram enclosing the erroneous terms we see that the lines form two sides of an isoceles triangle and that the apex of the triangle is the original erroneous term (0·3984). Next we examine the *size* of the errors in the various terms inside the triangle, or as it is often called "the error fan". The error in the original value of $\sin(0.4)$ was −0·0090 (since $0.3894 − 0.3984 = −0.0090$) and this propagates errors as shown

$f(x)$			−0·0090	
Δf		−0·0090		+0·0090
$\Delta^2 f$	−0·0090		+0·0180		−0·0090
$\Delta^3 f$	−0·0090	+0·0270	−0·0270	+0·0090

The striking features of this error fan are that every element is a multiple of the original error and the first erroneous element in each row is equal to the original error. That these features are not accidental is easily seen for suppose that we have a function,

Detection of Errors

$g(x)$, which is zero everywhere. The difference table for such a function is obviously composed entirely of zero entries. We now make a deliberate mistake and put $g(x) = 1$ at just one point. The difference table of $g(x)$ in the neighbourhood of the error is shown below

$$
\begin{array}{llllllllll}
g(x)=0 & & 0 & 0 & 0 & 1 & 0 & 0 & 0 & 0 \\
\Delta g = & & 0 & 0 & 0 & 1 & -1 & 0 & 0 & 0 \\
\Delta^2 g = & & & 0 & 0 & 1 & -2 & 1 & 0 & 0 \\
\Delta^3 g = & & & & 0 & 1 & -3 & 3 & -1 & 0 \\
\Delta^4 g = & & & & & 1 & -4 & 6 & -4 & 1 \\
\end{array}
$$

The numbers in the table are well-known: the binomial coefficients. Thus the members in the error fan in the Δ^4 row are just the coefficients of $x^0, x^1, \ldots, x^4$ in the expansion of $(1-x)^4$ since

$$(1-x)^4 = 1 - 4x + 6x^2 - 4x^3 + x^4.$$

This result is a special case of the following theorem, which we shall not prove here but which can be proved quite easily by using simple properties of the binomial coefficients.

THEOREM. If a table of values of a function $f(x)$ contains a single error of size h then the line $\Delta^k f$ contains $(k+1)$ consecutive terms in error and the error in the $(n+1)$ st term is equal to

$$(-1)^n {}_k C_n h \qquad n = 0, 1, \ldots, k$$

where ${}_k C_n = k!/(n!(n-k)!)$ is the Binomial coefficient.

The theorem tells us, as a bonus, the effect upon the difference table of rounding errors. A rounding error in $f(x)$ is multiplied by 2 in the worst case in $\Delta^2 f$, by 3 in the worst case in $\Delta^3 f$, by 6 in $\Delta^4 f$ and so on. So in Example 5.1 the rounding error in $f(x)$ is $\frac{1}{2} \times 10^{-4}$; in $\Delta^3 f$ this could be as big as $\frac{3}{2} \times 10^{-4} = 0 \cdot 00015$, even if we consider only the rounding error in a single value of $f(x)$—and in practice the error in $\Delta^3 f$ may be much worse than this because of the combined effect of several rounding errors. It is clear that care must be exercised in making use of higher order differences. An earlier instance of his problem was seen in Example 4.8.

We now have sufficient facts to enable us to attempt to detect, and possibly even to correct errors in a table of values of a

Numerical Analysis

function at equally spaced points. On the basis of the discussion and examples above our procedure should be:

(1) form a difference table of the values of the function; do not continue beyond the point where rounding errors are comparable in size to the differences;
(2) see if any pattern can be detected in most of the terms in some row $\Delta^k f$;
(3) Mark the terms in $\Delta^k f$ where the pattern seems to be missing, if there are $(k+1)$ terms in consecutive places we have probably detected the error;
(4) draw the "error fan" and so pick out the original erroneous value;
(5) See what change in the erroneous value would restore the pattern in $\Delta^k f$; if such a number can be found we can correct the error.

By way of illustration let us see an application of this procedure in practice.

Example 5.3. The values of a function at 8 equally spaced points are given below. The table contains an error. Find which value is in error and estimate its true value.

Solution

$f(x) =$	712	784	857	929	1010	1073	1144	1216
$f =$		72	73	72	81	63	71	72
$\Delta^2 f =$			+1	−1	+9	−18	+8	+1

We stop at the $\Delta^2 f$ line since we are already down to values which are affected by rounding errors in the last digit of $f(x)$. In this case the three erroneous terms in the $\Delta^2 f$ line are obviously +9, −18, +8 and we can draw the error fan which picks out 1010 as the original value in error. On the assumption that the terms in the Δf line should be 71, 72 or 73 we see that the error, based on the 81 or 63 is 10, 9 or 8. Similarly if the $\Delta^2 f$ line should consist of values +1, 0 or −1 the error, based on the values within the fan is somewhere between 7 and 10. Our estimate then, based on

Detection of Errors

the mean of these values is that the error is about 8 or 9 and that the value 1010 should be 1002 or 1001. (In fact the value should be 1001 since the values are $10^5(\log_{10} x - 0\cdot 47)$ for $x = 3(0\cdot005)3\cdot035$ where the values of the logarithm are taken from a set of 5-figure tables.)

Detection of the pattern in one of the lines $\Delta^k f$ will not always be as simple as in this case and if the tabulated function and its derivatives are not reasonably smooth there may be no simple pattern to detect. As an example of such a case consider the function, $f(x)$, defined by

$f(n) =$ the n-th prime number if n is an integer

$= 0$ if n is not an integer.

Then the difference table $f(n)$ for $n = 30(1)37$ is

$f(n) =$	113	127	131	137	139	149	151	157
$\Delta f =$		+14	+4	+6	+2	+10	+2	+6
$\Delta^2 f =$			−10	+2	−4	+8	−8	+4

and the only pattern is that all values in Δf, $\Delta^2 f$ are even. An error of even size provided that it was not large enough to destroy the monotonicity of $f(n)$ would not be detected unless we discovered the rule defining $f(n)$.

Provided these reservations are kept in mind the difference table can often be used very effectively in the detection of errors.

5.1. Multiple Errors

If the table of values contain two or more terms in error detection of the erroneous terms is incomparably more difficult; for each such term will generate an error fan and unless these terms are well separated in the table the error fans will overlap producing a complicated error pattern in the later rows. In the next example we look at two of the simplest cases.

Example 5.4. Study the error pattern generated by (i) two consecutive errors of $+1$; (ii) consecutive errors of $+1$ and -1.

Solution (i) We write down the error pattern in $f(x)$ itself and then form the difference table:

$f(x) =$	0	0	0	0	1	1	0	0	0	0
$\Delta f =$		0	0	0	1	0	-1	0	0	0
$\Delta^2 f =$			0	0	1	-1	-1	1	0	0
$\Delta^3 f =$				0	1	-2	0	2	-1	0
$\Delta^4 f =$					1	-3	2	2	-3	1

—and we see that the errors grow rather slowly.

(ii) Similarly:

$f(x) =$	0	0	0	0	1	-1	0	0	0	0
$\Delta f =$		0	0	0	1	-2	1	0	0	0
$\Delta^2 f =$			0	0	1	-3	3	-1	0	0
$\Delta^3 f =$				0	1	-4	6	-4	1	0

—and we see that in this case the errors in the $\Delta^k f$ are the same as those in the $\Delta^{(k+1)} f$ row of a function containing a single erroneous term with an error of one unit.

In the special case where we have two widely separated errors we may be able to detect the presence of two distinct error fans. In the case of a table of logarithms, as Example 5.3 shows, two errors separated by, say, 10 or more places in the table should be quite easy to detect, but this is mainly because of the slow growth rate of the logarithm.

If we now consider the case of a function $f(x)$ which has been tabulated to, say, 4 d.p. and ask what will be the effects on the difference table of the rounding errors in each value of $f(x)$ we realize that the effects will be unpredictable in detail and the most we can hope to do is to give an upper bound for the effect of these errors in the terms in each row $\Delta^k f$ for $k = 1, 2, \ldots$. We therefore examine the maximum rate at which combinations of errors can grow in a difference table.

Suppose that we have a table of values of a function $f(x)$ and that the maximum rounding error in any one value is ε in absolute value. Then since the line Δf is formed by taking the difference of two values in the $f(x)$ line the maximum rounding error in any value in the Δf line is 2ε. By the same argument the maximum rounding error in any value in the $\Delta^2 f$ line is 4ε, in the $\Delta^3 f$ line 8ε, and so on. Thus we have established:

Detection of Errors

THEOREM. If the maximum rounding error in any value of $f(x)$ is ε then the maximum error in any value in the line $\Delta^k f$ is $2^k \varepsilon$.

This upper bound, $2^k \varepsilon$, given by the theorem is a best-possible result for consider the case of a function $f(x)$ tabulated at a set of points with rounding errors $+\varepsilon, -\varepsilon$ alternately. The error pattern for $f(x)$ is

$$\ldots + \varepsilon - \varepsilon + \varepsilon - \varepsilon + \varepsilon - \varepsilon + \varepsilon - \varepsilon \ldots$$

and so the error pattern of the difference table is

$$\Delta f = \ldots -2\varepsilon + 2\varepsilon - 2\varepsilon + 2\varepsilon - 2\varepsilon + 2\varepsilon - 2\varepsilon \ldots$$
$$\Delta^2 f = \ldots + 4\varepsilon - 4\varepsilon + 4\varepsilon - 4\varepsilon + 4\varepsilon - 4\varepsilon \ldots$$
$$\Delta^3 f = \ldots - 8\varepsilon + 8\varepsilon - 8\varepsilon + 8\varepsilon - 8\varepsilon \ldots$$

and so on.

This result shows us that it is pointless carrying the difference table too far. So, in Example 5.1 we had a set of values of $\sin(x)$ accurate to 4 d.p. The rounding errors were therefore up to $\frac{1}{2} \times 10^{-4}$ in absolute value. In the line $\Delta^4 f$ the rounding error, from the theorem, might be as high as 8×10^{-4} and this is already larger than the values obtained by forming the fourth differences of $\sin(x)$ at the given interval. It was therefore wise not to use $\Delta^4 f$.

Problems on Chapter 5

(1) The values below are taken from a table of square roots for $x = 2 \cdot 0(0 \cdot 1) 2 \cdot 7$; one of the values is in error. Detect and correct the erroneous term and then compare your result with a table of square roots.

1·4142   1·4491   1·4832   1·5160   1·5492
1·5811   1·6125   1·6432.

(2) The values of a certain cubic have been tabulated at a set of equally spaced points. The values are printed as:

−3·369   −2·672   −1·903   −1·056   −0·135
+0·896   +2·013   +3·232   +4·559

Show that there is an error in this table; locate the erroneous term and correct it.

(3) Find the cubic which fits the corrected data in the example above given that it has been tabulated at $x = 1 \cdot 1(0 \cdot 1) 1 \cdot 9$.

Chapter 6

Numerical Integration

In scientific work it is often necessary to evaluate integrals, in one or more dimensions, which either cannot be evaluated at all by analytical methods or for which the analytical solution is very involved. In such cases we will have to use numerical methods and we are fortunate in that simple, reliable methods exist which will deal with most of the integrals that we are liable to encounter.

6.1. The Trapezium Rule

Suppose that we wish to evaluate

$$\int_a^b f(x)\,dx$$

where a, b are fixed, finite, given numbers, $b > a$ and $f(x)$ is a given function of the single variable which is finite and continuous throughout the interval $a \leq x \leq b$. If we draw the curve $y = f(x)$ then the value of the integral is the same as the area bounded by the lines $y = 0$, $x = a$, $x = b$ and the arc of the curve $y = f(x)$ between $x = a$ and $x = b$, as shown in the diagram

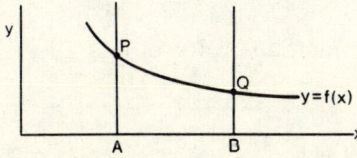

where A and B are the points $(a, 0)$ and $(b, 0)$ respectively. If P is the point $(a, f(a))$ and Q is the point $(b, f(b))$ then as a first, rather crude, approximation to the area we can take the area of

Numerical Analysis

the trapezium $APQB$. We therefore take

$$\int_a^b f(x)\,dx \doteqdot \tfrac{1}{2}(b-a)(f(a)+f(b)) \tag{6.1}$$

In practice we would normally divide the interval $\langle a, b\rangle$ into sub-intervals and apply the approximation formula (6.1) to the integral over each sub-interval in turn. Thus if we divide $\langle a, b\rangle$ into n sub-intervals by taking $(n+1)$ points $x_0, x_1, \ldots, x_n$ where $a = x_0 < x_1 < x_2 \ldots < x_n = b$ then we have

$$\int_a^b f(x)\,dx = \sum_{i=0}^{n-1} \int_{x_i}^{x_{i+1}} f(x)\,dx \tag{6.2}$$

and, as in (6.1), we take

$$\int_{x_i}^{x_{i+1}} f(x)\,dx \doteqdot \tfrac{1}{2}(x_{i+1}-x_i)(f(x_i)+f(x_{i+1})) \tag{6.3}$$

The lengths of the sub-intervals need not all be the same though they usually are. In the case where they are the same we have

$$x_{i+1} - x_i = h \text{ (say)}$$

so that

$$b - a = nh$$

and (6.2) and (6.3) combined then give us:

$$\int_a^b f(x)\,dx = \int_{x_0}^{x_0+nh} f(x)\,dx$$

$$\doteqdot \frac{h}{2}\Big[f(x_0) + 2f(x_1) + 2f(x_2) + \ldots + 2f(x_{n-1}) + f(x_n)\Big] \tag{6.4}$$

and this is called the Trapezium Rule.

Example 6.1. Estimate the value of $\log_e 2$ by a Trapezium Rule approximation to

$$\int_1^2 \frac{dx}{x}, \quad \text{taking} \quad h = \frac{1}{3}$$

Solution. $f(x) = 1/x$ in this case and since $h = \tfrac{1}{3}$ we shall require the values of $f(x)$ at $x = 1, \frac{4}{3}, \frac{5}{3}$ and 2, viz.

x	$f(x)$
1	1·0000
$\frac{4}{3}$	0·7500
$\frac{5}{3}$	0·6000
2	0·5000

and the Trapezium Rule gives

$$\log_e 2 \doteqdot \frac{1}{6}[1\cdot 0000 + 1\cdot 5000 + 1\cdot 2000 + 0\cdot 5000]$$

i.e. $\log_e 2 \doteqdot 0\cdot 7000$

Although we have carried out the work to 4 d.p. there are no *rounding* errors in this case. There is however a *truncation* error since we are approximating an integral by a sum.

Now $\log_e 2 = 0\cdot 69315$ so our relative error is about 1%. What happens if we use a smaller value of h? Will the relative error decrease and, if so, by what factor? The next example throws some light on this matter.

Example 6.2. Repeat the previous example but take $h = \frac{1}{6}$. What is the relative error?

Solution. We now need the values of $f(x)$ at $x = \frac{7}{6}, \frac{3}{2}$ and $\frac{11}{6}$ in addition to the previous values, viz. (to 4 d.p.)

x	$f(x)$
1	1·0000
$\frac{7}{6}$	0·8571
$\frac{4}{3}$	0·7500
$\frac{3}{2}$	0·6667
$\frac{5}{3}$	0·6000
$\frac{11}{6}$	0·5455
2	0·5000

and so

$$\log_e 2 \doteqdot \frac{1}{12}[1\cdot 000 + 1\cdot 7142 + 1\cdot 5000 + 1\cdot 3334 + 1\cdot 2000$$
$$+ 1\cdot 0910 + 0\cdot 5000]$$

i.e. $\log_e 2 \doteqdot 0\cdot 6949$

Numerical Analysis

The relative error is now approximately

$$\left(\frac{0\cdot 0017}{0\cdot 6932}\right) 100\% = \frac{1}{4}\%$$

If we now halve the interval size again (i.e. take $h = \frac{1}{12}$) and repeat the calculation we obtain

$$\log_e 2 \doteqdot 0\cdot 6936$$

and the relative error is now only about $\frac{1}{16}\%$. These three results indicate that the truncation error when we use the Trapezium Rule with an interval size h is proportional to h^2 and this is indeed the case, as we shall see in section 6.6.

6.2. The Midpoint Rule

This method is fundamentally similar to the Trapezium Rule. As before we suppose that we wish to evaluate $\int_a^b f(x)\,dx$ and draw the diagram as in section 6.1.

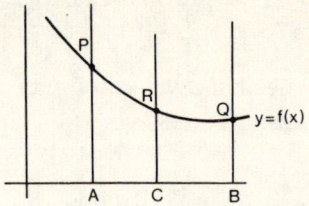

but, in addition we mark the points $C((a+b)/2, 0)$ which is the midpoint of AB and $R((a+b)/2, f((a+b)/2))$ where the ordinate through C meets the curve $y = f(x)$. We then take the area of the rectangle based on AB and of height CR to be an approximation to the area under the curve between A and B, i.e. to the value of the integral. Thus our approximation is:

$$\int_a^b f(x)\,dx \doteqdot (b-a)f\left(\frac{a+b}{2}\right) \qquad (6.4)$$

This is the Midpoint Rule for a single interval.

Example 6.3. Use the Midpoint Rule based on single interval to estimate $\log_e 2$.

Solution. As in Example 6.1.

$$\log_e 2 = \int_1^2 \frac{dx}{x}$$

$$\doteq (1)\left(\frac{2}{3}\right) = 0\cdot 6667 \text{ to 4 d.p.}$$

The relative error is about 4%.

We can easily adapt (6.4) to cover the case of several intervals. If, as before, we subdivide $\langle a\, b \rangle$ so that

$$a = x_0 < x_1 < \ldots < x_n = b$$

we have $b - a = nh$ and so

$$\int_a^b f(x)\, dx = \int_{x_0}^{x_0+nh} f(x)\, dx \doteq (x_{i+1} - x_i) \sum_{i=0}^{n-1} f\left(\frac{x_i + x_{i+1}}{2}\right)$$

or, if the points $x_0, x_1, \ldots, x_n$ are equally spaced

$$\int_{x_0}^{x_0+nh} f(x)\, dx \doteq h \sum_{k=0}^{n-1} f(x_0 + (k + \tfrac{1}{2})h) \tag{6.5}$$

Example 6.4. Use the Midpoint Rule with $h = \tfrac{1}{3}$ to estimate $\log_e 2$.

Solution. $f(x) = (1/x)$ and, from (6.5), we shall require $f(\tfrac{7}{6})$, $f(\tfrac{3}{2})$, $f(\tfrac{11}{6})$. Then

$$\log_e 2 = \int_1^2 \frac{dx}{x} \doteq \frac{1}{3}\left(\frac{6}{7} + \frac{2}{3} + \frac{6}{11}\right)$$

$$= \frac{1}{3}(0\cdot 8571 + 0\cdot 6667 + 0\cdot 5455) \text{ to 4 d.p.}$$

$$\doteq 0\cdot 6898$$

The relative error is now about $\tfrac{1}{2}$% which is better than the Trapezium Rule and the error is of the opposite sign. A glance at the diagram makes it clear why, in the case of the curve $y = (1/x)$, the Trapezium Rule *overestimates* the value of the integral whereas the Midpoint Rule *underestimates* it. In the next section we shall estimate the truncation errors of these two methods and we shall see that these will often be of opposite signs and that of the Midpoint Rule will usually be the smaller in absolute value.

6.3. Truncation Errors in the Trapezium and Midpoint Rules

In order to estimate the truncation error we make use of the Taylor Series expansion of $f(x)$. It is proved in courses in Pure Mathematics that if $f(x)$ and all its derivatives exist and are continuous in the interval $\langle 0, x \rangle$ then for any $n \geq 1$,

$$f(x) = f(0) + xf'(0) + \frac{x^2}{2!}f''(0) + \frac{x^3}{3!}f'''(0) + \ldots + \frac{x^{n-1}}{(n-1)!}f^{n-1}(0)$$

$$+ \frac{x^n}{n!}f^n(\xi)$$

where $\qquad 0 < \xi < x$

In general we have no idea what value ξ has for any given value of x but in the special case when $f(x)$ is a polynomial of degree k $f^n(x)$ is identically zero if $n > k$, the value of ξ is irrelevant and the expansion above is a simple identity in powers of x.

For the applications in this section we choose $n = 2$ in the expansion so that:

$$f(x) = f(0) + xf'(0) + \frac{x^2}{2!}f''(\xi) \qquad (6.6)$$

Now we take the Trapezium Rule in the form:

$$\int_0^h f(x)\,dx \doteqdot \frac{h}{2}(f(0) + f(h)) \qquad (6.7)$$

Replacing $f(x)$ by the expression in (6.6) gives

$$\int_0^h f(x)\,dx = \int_0^h \left(f(0) + xf'(0) + \frac{x^2}{2!}f''(\xi_1)\right)dx \qquad (6.8)$$

where $\qquad 0 < \xi_1 < h$

and we treat $f''(\xi_1)$ as constant—it is not since ξ_1 is a function of x, but we are making the tacit assumption that the interval $\langle 0, h \rangle$ is sufficiently small that $f''(x)$ doesn't change very much over the interval. Carrying out the integration (6.8) gives

$$\int_0^h f(x)\,dx = hf(0) + \frac{h^2}{2}f'(0) + \frac{h^3}{6}f''(\xi_1) \qquad (6.9)$$

Numerical Integration

The Trapezium Rule gives

$$\int_0^h f(x)\,dx \doteqdot \frac{h}{2}(f(0)+f(h))$$

$$= \frac{h}{2}\left(2f(0) + hf'(0) + \frac{h^2}{2}f''(\xi_2)\right)$$

$$= hf(0) + \frac{h^2}{2}f'(0) + \frac{h^3}{4}f''(\xi_2) \qquad (6.10)$$

and comparison of (6.9) and (6.10) combined with the assumption that $f''(\xi_2) \doteqdot f''(\xi_1)$ leads to the error estimate for the Trapezium Rule, viz.

$$\int_0^h f(x)\,dx - \frac{h}{2}(f(0)+f(h)) = -\frac{h^3}{12}f''(\xi_1) \qquad (6.11)$$

where $\qquad 0 < \xi_1 < h$

Next we take the Midpoint Rule in the form:

$$\int_0^h f(x)\,dx \doteqdot hf\left(\frac{h}{2}\right) \qquad (6.12)$$

and Taylor's Expansion leads to

$$hf\left(\frac{h}{2}\right) = hf(0) + \frac{h^2}{2}f'(0) + \frac{h^3}{8}f''(\xi_3)) \qquad (6.13)$$

Comparison of (6.9) and (6.13) gives the error in the Midpoint Rule as

$$\int_0^h f(x)\,dx - hf\left(\frac{h}{2}\right) = +\frac{h^3}{24}f''(\xi_3) \qquad (6.14)$$

for some number ξ_3 satisfying $0 < \xi_3 < h$ and we see from (6.11) and (6.14) that

(i) the errors will be of opposite sign if $f''(x)$ doesn't change sign in the interval $\langle 0, h \rangle$ and
(ii) that the magnitude of the error in the Midpoint Rule is about half of that in the Trapezium Rule.

Note that these error estimates are for a *single application* of the Trapezium and Midpoint Rules. In 6.6 we shall discuss the errors when we apply these rules over several intervals.

6.4. Simpson's Rule

In both the Trapezium Rule and the Midpoint Rule our approximation to the area under the curve is obtained by replacing the arc of the curve between $x = a$ and $x = b$ by a straight line. In the former case the line joined the points at the ends of the arc; in the latter case it passes through the mid-point of the arc, parallel to the x-axis. A natural extension of this idea is to replace the arc by a curve of the second degree in x which passes through the two end points and the mid-point of the arc. Such a curve is a parabola and has an equation of the form

$$y = Ax^2 + Bx + C$$

The algebra is simplified if we take $a = -h$, $b = +h$ so that $\frac{1}{2}(a+b) = 0$. We then wish to choose A, B, C so that the parabola passes through the points $(-h, f(-h))$, $(0, f(0))$, $(h, f(h))$. The equations defining A, B, C are therefore

$$Ah^2 - Bh + C = f(-h)$$
$$C = f(0)$$
$$Ah^2 + Bh + C = f(h)$$

so that:
$$A = \frac{f(-h) + f(h) - 2f(0)}{2h^2}$$

$$B = \frac{f(h) - f(-h)}{2h}$$

$$C = f(0)$$

The parabola so defined coincides with the curve $y = f(x)$ at the three points where $x = -h, 0, +h$. We may therefore say that

$$\int_{-h}^{h} f(x)\, dx \doteqdot \int_{-h}^{h} (Ax^2 + Bx + C)\, dx = \frac{2}{3} Ah^3 + 2Ch$$

and substitution of the values of A, B, C gives

$$\int_{-h}^{h} f(x)\, dx \doteqdot \frac{h}{3}(f(-h) + 4f(0) + f(h)) \qquad (6.15)$$

which is *Simpson's Rule for a single interval*.

If we wish to use Simpson's Rule over the interval $\langle a, b \rangle$ rather

Numerical Integration

than the interval $\langle -h, +h \rangle$ it takes the form

$$\int_a^b f(x)\,dx \doteqdot \frac{|b-a|}{6}\left(f(a) + 4f\left(\frac{a+b}{2}\right) + f(b)\right)$$

Thus, in particular:

$$\int_0^{2h} f(x)\,dx \doteqdot \frac{h}{3}(f(0) + 4f(h) + f(2h)) \qquad (6.16)$$

and

$$\int_{2h}^{4h} f(x)\,dx \doteqdot \frac{h}{3}(f(2h) + 4f(3h) + f(4h)) \qquad (6.17)$$

and so, adding (6.16) and (6.17)

$$\int_0^{4h} f(x)\,dx \doteqdot \frac{h}{3}(f(0) + 4f(h) + 2f(2h) + 4f(3h) + f(4h))$$

Continuing in this way it is easy to see that the general form of Simpson's Rule is

$$\int_0^{2nh} f(x)\,dx \doteqdot \frac{h}{3}[(f(0) + f(2nh) + 4(f(h) + f(3h) + \ldots + f(2n-1)h)$$
$$+ 2(f(2h) + f(4h) + \ldots + f((2n-2)h))]$$

This formula may appear to be rather involved but it is really very simple. The coefficients of $f(0)$, $f(h)$, $f(2h)$, ... are

$$1, 4, 2, 4, 2, 4, 2, \ldots, 2, 4, 1$$

Example 6.5. Use Simpson's Rule on 3 points to estimate the value of $\log_e 2$.

Solution. We approximate to $\int_1^2 (dx/x)$ at $x = 1 \cdot 0$, $1 \cdot 5$ and $2 \cdot 0$ so that in this case $h = \frac{1}{2}$. Working to 4 d.p.

x	$f(x)$
1·0	1·0000
1·5	0·6667
2·0	0·5000

so that
$$\log_e 2 \doteqdot \frac{1}{6}(1 \cdot 0000 + 4(0 \cdot 6667) + 0 \cdot 5000)$$

i.e.
$$\log_e 2 \doteqdot 0 \cdot 6944$$

Numerical Analysis

The relative error is less than $\frac{1}{4}$%, although the result is only correct to 2 d.p. If we use Simpson's Rule on 5 points (i.e. $h = \frac{1}{4}$) instead of 3 the relative error is reduced to about $\frac{1}{70}$%, an improvement by a factor of approximately 16 (see Example 6.6 at the end of this Chapter). This indicates that the truncation error in Simpson's Rule is proportional to h^4. In the next two sections we shall study the truncation error in Simpson's Rule over a single interval and then, more generally, the truncation error when we use any of our numerical integration methods over several intervals.

6.5. Truncation Error in Simpson's Rule

We take Simpson's Rule in the form given by (6.15) ar estimate the truncation error over the interval $\langle -h, h \rangle$.
By Taylor's Theorem

$$f(x) = f(0) + xf'(0) + \frac{x^2}{2!}f''(0) + \frac{x^3}{3!}f'''(0) + \frac{x^4}{4!}f''''(\alpha) \quad (6.18)$$

where

$$0 < \alpha < x.$$

Therefore $\int_{-h}^{h} f(x)\,dx = 2hf(0) + \frac{h^3}{3}f''(0) + \frac{h^5}{60}f''''(\beta) \quad (6.19)$

for some number β in $\langle -h, h \rangle$.
Now, from (6.15), Simpson's Rule is

$$\int_{-h}^{h} f(x)\,dx \doteqdot \frac{h}{3}\{f(-h) + 4f(0) + f(h)\}$$

and, from Taylor's Theorem

$$f(-h) = f(0) - hf'(0) + \frac{h^2}{2!}f''(0) - \frac{h^3}{3!}f'''(0) + \frac{h^4}{4!}f''''(y)$$

and $f(h) = f(0) + hf'(0) + \frac{h^2}{2!}f''(0) + \frac{h^3}{3!}f'''(0) + \frac{h^4}{4!}f''''(\delta)$

where $-h < y < 0$ and $0 < \delta < h$

so that

$$\frac{h}{3}\{f(-h)+4f(0)+f(h)\} \doteqdot 2hf(0)+\frac{h^3}{3}f''(0)+\frac{h^5}{36}f''''(\varepsilon) \quad (6.20)$$

on the assumption that $f''''(y) \doteqdot f''''(\delta) \doteqdot f''''(\varepsilon)$.
If we now compare (6.19) and (6.20) we see that the truncation error in Simpson's Rule is approximately

$$\frac{h^5}{60}f''''(\beta)-\frac{h^5}{36}f''''(\varepsilon)$$

or, assuming that $f''''(\beta) = f''''(\varepsilon)$

$$\frac{-h^5}{90}f''''(\beta)$$

for some β in $\langle -h, h \rangle$.

This argument can be made completely rigorous; it is not necessary to assume that $f''''(\beta) = f''''(\varepsilon)$ but the proof is then more complicated and all we are really interested in at present is that the Truncation Error in a single application of Simpson's Rule is proportional to h^5 and that in the Midpoint Rule and the Trapezium Rule the Truncation Error in a single application is proportional to h^3.

6.6. Truncation Errors over Several Intervals

In Example 6.2 we saw that when we used the Trapezium Rule with $h = \frac{1}{6}$ the truncation error was reduced to about one quarter of what it was when $h = \frac{1}{3}$. This implies, as was pointed out before, that the truncation error in the Trapezium Rule is proportional to h^2. On the other hand we saw in section 6.3 that the truncation error in a single application of the Trapezium Rule is proportional to h^3. Is there some contradiction here?

The answer is that there is no contradiction. For suppose we apply the Trapezium Rule over a single interval $\langle 0, h \rangle$ to the evaluation of $\int_0^h f(x)\,dx$ then the truncation error, by (6.11) is

$$\frac{-h^3}{12} f''(\xi) \quad \text{where} \quad 0 < \xi < h.$$

If we now halve the interval size to $\tfrac{1}{2}h$ we have to apply the Trapezium Rule twice, viz.

$$\int_0^h f(x)\,dx = \int_0^{1/2h} f(x)\,dx + \int_{h/2}^h f(x)\,dx$$

and from (6.11) each of these applications will produce a truncation error, in fact:

$$\int_0^{h/2} f(x)\,dx = \frac{h}{4}\left(f(0) + f\!\left(\frac{h}{2}\right)\right) - \frac{(\tfrac{1}{2}h)^3}{12} f''(\xi_1)$$

and $$\int_{h/2}^h f(x)\,dx = \frac{h}{4}\left(f\!\left(\frac{h}{2}\right) + f(h)\right) - \frac{(\tfrac{1}{2}h)^3}{12} f''(\xi_2)$$

where $0 < \xi_1 < \tfrac{1}{2}h$ and $\tfrac{1}{2}h < \xi_2 < h$
so that, adding:

$$\int_0^h f(x)\,dx = \frac{h}{4}\left(f(0) + 2f\!\left(\frac{h}{2}\right) + f(h)\right) - (\tfrac{1}{2}h)^3(f''(\xi_1) + f''(\xi_2))$$

On the assumption that $f''(\xi_1) = f''(\xi_2)$ the *total* truncation error is approximately

$$\frac{(\tfrac{1}{2}h)^3}{12} (2f''(\xi_1)) = \left(\frac{1}{4}\right)\left(\frac{h^3}{12}\right) f''(\xi_1)$$

i.e. is one quarter, *not* one eighth, of its previous value.

The explanation, in words, is quite simple: when we halve the interval size we introduce two truncation errors each of which is approximately one eighth of the original truncation error so that the new total truncation error is about one quarter of its original value.

There is nothing special about *halving* the interval size. If we divided the interval into n parts we would have n truncation errors each of which would be approximately equal to $1/n^3$ times the original truncation error. Overall therefore the truncation error would be reduced by a factor of about n^2, not n^3.

Similar arguments can be applied to study the effect on the truncation error when the interval size is reduced by a factor of n

in the cases of the Midpoint Rule and Simpson's Rule. The answers are that the errors are reduced by factors of about n^2 and n^4 respectively.

If we use any of these methods of numerical integration on a computer to evaluate say

$$\int_1^2 \frac{dx}{x} \text{ with interval sizes of } h = 1, \tfrac{1}{2}, \tfrac{1}{4}, \tfrac{1}{8}, \ldots$$

we know that the truncation errors will decrease rapidly but does this mean that our answers will eventually become as accurate as we wish? Unfortunately they will not, because in addition to truncation errors we also have rounding errors which are insignificant in the example above when $h = 1$ but which become increasingly significant as h becomes smaller, because we are adding together more and more terms each of which involves a rounding error. Eventually we reach a value of h for which the rounding errors are larger than the truncation error and from this point onwards as we decrease h our answer will get steadily worse instead of better. Thus, on any given computer, there is a limit to the accuracy that can be obtained in attempting numerical evaluation of integrals and this obviously depends on the effective word-size of the machine; furthermore, and this is the really important point, any attempt to obtain greater accuracy by reducing the interval size still further will be self-defeating for accumulated rounding errors will *decrease* the accuracy, not increase it. This phenomenon is illustrated by:

Example 6.6. Write a program to find the value of $\log_e 2$ by using the Trapezium Rule to evaluate $\int_1^2 (dx/x)$. Start with $h = \tfrac{1}{2}$, then $h = \tfrac{1}{4}, \tfrac{1}{8}, \tfrac{1}{16}, \ldots, \tfrac{1}{4096}$. Print the 12 approximate values for $\log_e 2$ so found. Compare your results with the known value of $\log_e 2$ (0·69314718 to 8 d.p.)

Solution. (This problem was originally set for first-year students at Cardiff using Fortran on the ICL 4-70; the results on other machines may be different. If a machine of large word size were used it would be necessary to reduce h still further before the phenomenon would be seen.)

The results on the 4-70 are shown. We take $h = (1/N)$ and

print N, the value obtained for $\log_e 2$ and the absolute value of the error. Results are given to 6 d.p.

N	Value	$\|Error\|$
2	0·708333	0·015186
4	0·697024	0·003877
8	0·694122	0·000975
16	0·693391	0·000244
32	0·693208	0·000051
64	0·693161	0·000014
128	0·693148	0·000001
256	0·693142	0·000005
512	0·693134	0·000013
1024	0·693120	0·000027
2048	0·693092	0·000055
4096	0·693034	0·000113

We note that the error is a minimum at $h = \frac{1}{128}$. From that point onwards the accumulated rounding error dominates the truncation error.

Problems on Chapter 6

(1) Evaluate $\int_0^1 x^{1/2} dx$ by (a) the Trapezium Rule, (b) Simpson's Rule, taking $h = \frac{1}{4}$ in each case. Compare your results with the exact value to 4 d.p.

(2) Evaluate $\int_1^2 x^{1/2} dx$ by Simpson's Rule with $h = \frac{1}{4}$ and compare with the exact result. Why is the error so much smaller than in case (b) of Question 1?

(3) Use Simpson's Rule with a step size of $\pi/8$ to find an approximate value for

$$\int_0^{\pi/2} \sqrt{1 + \sin x}\, dx.$$ Work to 4 d.p.

(Cardiff Part 1, 1976)

(4) Apply Simpson's Rule with $h = \frac{1}{6}$ to the evaluation of $\int_0^1 \sin \pi x\, dx$ and then, by direct evaluation of the integral show that $\pi \doteq (18(4 - \sqrt{3}))/13$.

(Cardiff Part 1, 1974)

(5) Let I_1, I_2 and I_3 be the results of evaluating $\int_{-h}^{h} f(x)\, dx$ by the Trapezium Rule, Midpoint Rule and Simpson's Rule respectively. Show that $3I_3 = I_1 + 2I_2$.

Chapter 7

The Solution of Non-Linear Equations

In this chapter we consider the problem of solving an equation of the type $f(x)=0$ where $f(x)$ is a given *non-linear* function. A function $g(x)$ is said to be *linear* if it can be written in the form

$$g(x) = ax + b$$

where a, b are constants, otherwise it is non-linear. Thus x^2, $3x^2 - 2x + 7$, $x \sin x$, $2e^x + \cos x + x + 3$ are all examples of non-linear functions. In general there is no analytical method for solving non-linear equations and so we must use numerical methods. In the sections that follow we shall develop and study five such methods. These five methods fall into two classes:
(1) two-point methods and
(2) one-point methods;
and we shall discuss them in that order.

7.1. Two-point Methods. (1) The Method of Bisection

Suppose that we wish to solve

$$f(x) = 0 \qquad (7.1)$$

and that we have found two approximate values for the solution x_1, x_2 such that $f(x_1)$ and $f(x_2)$ have opposite signs, i.e. such that

$$f(x_1)f(x_2) < 0 \qquad (7.2)$$

It follows that assuming $f(x)$ to be continuous over $\langle x_1, x_2 \rangle$, there is a solution to (7.1) somewhere in the interval $\langle x_1, x_2 \rangle$. We therefore have the problem of how to choose a value x_3 such that

Numerical Analysis

$x_1 < x_3 < x_2$ and $f(x_3) = 0$. The simplest method, from a computational point of view is to take

$$x_3 = \tfrac{1}{2}(x_1 + x_2) \tag{7.3}$$

We now evaluate $f(x_3)$; if $f(x_1)f(x_3) < 0$ we choose a new point

$$x_4 = \tfrac{1}{2}(x_1 + x_3) \tag{7.4}$$

whereas if $f(x_1)f(x_3) > 0$ then $f(x_2)f(x_3) < 0$ and we choose

$$x_4 = \tfrac{1}{2}(x_2 + x_3) \tag{7.5}$$

and so on. At every stage we have had two points x_i, x_j such that $f(x_i)f(x_j) < 0$ and we choose the next point to be

$$x_e = \tfrac{1}{2}(x_i + x_j) \tag{7.6}$$

and use this point and whichever of x_i, x_j causes $f(x)$ to have the opposite sign to $f(x_e)$ as the two points for the following stage. The process terminates when we reach a point x_n such that $|f(x_n)|$ is sufficiently small.

This method is known as the *Method of Bisection*.

Example 7.1. Find a solution of the equation

$$\sin(x) - \tfrac{1}{2}x = 0$$

in the interval $\langle \tfrac{1}{2}\pi, \pi \rangle$.

Solution. We first note that, taking $x_1 = \tfrac{1}{2}\pi$, $x_2 = \pi$ and writing $f(x) = \sin(x) - \tfrac{1}{2}x$ we have $f(\tfrac{1}{2}\pi) = 1 - \tfrac{1}{4}\pi > 0$, $f(\pi) = -\tfrac{1}{2}\pi < 0$ so that there is certainly at least one solution of the equation in the interval. We simplify the calculations by taking $x_1 = 1.5$, $x_2 = 3.0$, then on a computer or a desk machine we find (to 5 d.p.)

$x_3 = 2.25$ $\qquad f(x_3) = -0.34693$
$x_4 = 1.875$ $\qquad f(x_4) = +0.01659$
$x_5 = 2.0625$ $\qquad f(x_5) = -0.14972$
$\vdots \qquad \vdots$
$x_{10} = 1.89258$ $\qquad f(x_{10}) = +0.00238$
$\vdots \qquad \vdots$
$x_{19} = 1.89550$ $\qquad f(x_{19}) = -0.00000(1)$
$x_{20} = 1.89549$ $\qquad f(x_{20}) = +0.00000(2)$

The result is now correct to 5 d.p. but convergence has been slow: 18 iterations to reach the required accuracy. In the following sections we shall use methods which are a lot quicker and later we shall investigate the speed of convergence of the methods of this Chapter.

7.2. The Method of False Position

The method of bisection is very simple to use, easy to program for a computer and, as we shall see in 7.4, is certain to convergeto a solution of (7.1) but, as a study of Example 7.1 will show, it is unnecessarily slow in its convergence. Observe that $f(x_3) = -0.34693$ and $f(x_4) = +0.01659$; is it not apparent that the solution of (7.1) is probably closer to x_4 than it is to x_3? Under these circumstances it would surely be much more sensible to choose x_5 quite close to x_4 rather than midway between x_3 and x_4.

The method which we now describe takes advantage of just this kind of situation. Let us look at the problem geometrically. Suppose that we are trying to solve (7.1), that x_1, x_2 are approximate solutions and that $f(x_1)f(x_2) < 0$. In the diagram below let P be the point $(x_1, f(x_1))$ and let Q be the point $(x_2, f(x_2))$ and let the curve be $y = f(x)$. Let the chord PQ meet the x-axis in $R(x_3, 0)$ then we take x_3 as our next approximation to the solution of (7.1)

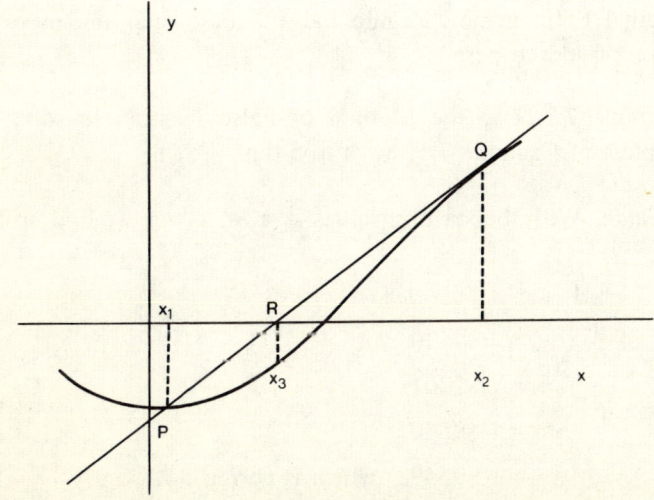

Numerical Analysis

The formula for x_3 is easily found. The equation of the chord PQ is

$$\frac{y-f(x_1)}{x-x_1} = \frac{y-f(x_2)}{x-x_2}$$

and this line meets the line $y=0$ where

$$x_3 = \frac{x_1 f(x_2) - x_2 f(x_1)}{f(x_2) - f(x_1)} \tag{7.7}$$

So, given two approximate values x_1, x_2 to the solution we can construct a new approximate value x_3, from (7.7) which, hopefully, will be closer to the true solution.

The question now arises: from x_1 and x_2 we have constructed a new approximation x_3. How should we construct the next approximation x_4? There are clearly two possibilities:

(i) to use x_1 and x_3;

or (ii) to use x_2 and x_3.

How do we decide? One reasonable way is as follows: evaluate $f(x_3)$ and choose x_1 and x_3 if $f(x_1)f(x_3) < 0$, otherwise choose x_2 and x_3 (since $f(x_2)f(x_3)$ will be negative). This is the *Method of False Position* which was used many centuries ago, under its original Latin name "Regulo Falso", to solve problems of the type considered here.

Example 7.2. Use the Method of False Position to solve the problem of Example 7.1; work to 5 d.p.

Solution. With the starting values $x_1 = \frac{1}{2}\pi$, $x_2 = \pi$ we find, using a computer:

$$x_3 = 1 \cdot 75960$$
$$x_4 = 1 \cdot 84420$$
$$x_5 = 1 \cdot 87701$$
$$\vdots$$
$$x_{10} = 1 \cdot 89540$$
$$\vdots$$
$$x_{13} = 1 \cdot 89549, \quad \text{which is correct to 5 d.p.}$$

Convergence has therefore occurred in 11 iterations compared to 18 required by the Method of Bisection.

7.3. The Secant Method

In the Secant Method we also start from two approximate solutions x_1, x_2 but we do not insist that $f(x_1)f(x_2)<0$. We again apply formula (7.7) to find x_3 but in order to find x_4 we use x_2 and x_3 *irrespective* of the signs of $f(x_2)$ and $f(x_3)$; similarly we use x_3 and x_4 to find x_5 and so on.

It is not immediately apparent whether this method will converge or not and, if it does so, whether convergence will be faster or slower than with the method of false position. If we were to re-work Example 7.1 again, with the same starting values we would find that convergence to the value 1·89549 occurs in 5 steps and this indicates that the Secant Method converges at least sometimes and that if it does converge it might converge quite quickly.

7.4. Convergence of the Two-point Methods

The examples of the previous sections of this chapter indicated that the Secant method, if it converges, is faster than either the Regulo Falso or Bisection method. Can we however guarantee that these three methods will always converge to a solution of (7.1)? We first consider the Bisection method.

THEOREM. If $f(x)$ is continuous and if $f(x) = 0$ has a root in the interval $\langle x_1, x_2 \rangle$ then the method of bisection converges to the root.

Proof. Let ρ be a root in $\langle x_1, x_2 \rangle$ and let $d = |x_2 - x_1|$, then ρ lies in an interval of length d. We now take $x - \frac{1}{2}(x_1 + x_2)$ so that $|x_3 - x_2| = |x_3 - x_1| = \frac{1}{2}d$. Since ρ lies in one of the intervals $\langle x_1, x_2 \rangle$, $\langle x_3, x_2 \rangle$ it follows that ρ now lies in an interval of length $\frac{1}{2}d$. Continuing in this way after n steps we will have an interval containing ρ of length $d/2^n$. Thus we will have a number, x_{n+2}, such that $|x_{n+2}-\rho|<d/2^n$. Therefore $\lim_{n\to\infty}|x_n-\rho|=0$ and it follows that as $n\to\infty$, $x_n\to\rho$, i.e. the sequence $x_3, x_4, x_5 \ldots$ generated by the method of bisection converges to the root.

The proof also shows how many iterations are required in order to ensure that the value of the root is found within a prescribed accuracy. For if we require x_k to be correct to m places of decimals we must have

$$|x_k-\rho|<\tfrac{1}{2}\times 10^{-m}$$

but

$$|x_k-\rho|\doteqdot\frac{d}{2^{k-2}}$$

and so we must have

$$\frac{d}{2^{k-2}}\doteqdot\frac{1}{2\times 10^m}$$

or

$$2^k = 8d\times 10^m \doteqdot 2^{(10m/3)+3}\times d$$

Thus if $d\doteqdot 1$ then k will need to be about $(10m/3)+3$.

In Example 7.1 $d=1\cdot 5$ and $m=5$ so that we would expect k to be given by

$$2^k\doteqdot 12\times 10^5 = 1\cdot 2\times 10^6 \doteqdot 2^{20}, \quad \text{i.e.} \quad k\doteqdot 20$$

which agrees with our observed result.

Next we examine the convergence of the Regulo Falso. Suppose that we have two initial estimates x_1, x_2 for the root ρ. Since the root lies between x_1 and x_2 we may suppose without loss of generality that $x_1=\rho-\varepsilon_1$, $x_2=\rho+\varepsilon_2$ where $\varepsilon_1, \varepsilon_2$ are positive. Then our next approximation to the root is given by (7.7), viz.

$$x_3 = \frac{x_1 f(x_2) - x_2 f(x_1)}{f(x_2) - f(x_1)}$$

or

$$x_3 = \rho - \frac{\varepsilon_1 f(x_2) + \varepsilon_2 f(x_1)}{f(x_2) - f(x_1)}$$

i.e.

$$x_3 = \rho - \varepsilon_3, \quad \text{say.}$$

The Solution of Non-Linear Equations

Now $f(x_1)$ and $f(x_2)$ are of opposite signs and so

$$|\varepsilon_3| = \frac{|\varepsilon_1 f(x_2) + \varepsilon_2 f(x_1)|}{|f(x_2) - f(x_1)|}$$

$$= \frac{|\varepsilon_1 f(x_2) + \varepsilon_2 f(x_1)|}{|f(x_2)| + |f(x_1)|} \tag{7.8}$$

Suppose that $|\varepsilon_1 f(x_2)| < |\varepsilon_2 f(x_1)|$ then (7.8) tells us that

$$|\varepsilon_3| < \frac{|\varepsilon_2| |f(x_1)|}{|f(x_2)| + |f(x_1)|} < k_2 |\varepsilon_2| \tag{7.9}$$

where $k_2 < 1$.

From (7.9) we see that the error at each stage of the iteration will decrease in the sense that it will be smaller than the larger of the two previous errors, but it may not be much smaller and in particular this is likely to be true when we are close to the root. We infer that convergence will occur but it may be slow, particularly in the later stages.

This argument can be made quite rigorous and it is indeed true that the Regulo Falso method converges, but slowly.

In the case of the Secant Method it can be shown that convergence to the root may not occur at all which is not too surprising since we have dropped the requirement that $f(x_1)f(x_2) < 0$. As a trivial illustration of this consider the function

$$f(x) = x^2 - 2$$

Clearly $f(x)$ has two zeros at $x = \pm\sqrt{2}$. Suppose we take $x_1 = -1$, $x_2 = +1$ then $f(-1) = f(1) = -1$ and the line joining $(-1, 1)$ to $(+1, 1)$ is parallel to the x-axis and our next estimate is $x_3 = \infty$. It is not hard to construct less extreme examples of this kind. If the secant method does converge however it can be proved that the number of iterations required will be only about 60% of the number of iterations required by the Regulo Falso.

7.5. One-point Methods. The Newton-Raphson Method

If, in the Secant Method, we let the two starting values x_1, x_2 become arbitrarily close we eventually replace the secant joining the points $P_1(x_1, f(x_1))$, $P_2(x_2, f(x_2))$ by the tangent at P_1. The method so obtained then becomes a "One-point method". If we take (7.7)

$$x_3 = \frac{x_1 f(x_2) - x_2 f(x_1)}{f(x_2) - f(x_1)}$$

and write $x_2 = x_1 + h$ then
$f(x_2) \doteq f(x_1) + hf'(x_1)$ and so

$$x_3 \doteq \frac{x_1(f(x_1) + hf'(x_1)) - (x_1 + h)f(x_1)}{hf'(x_1)}$$

i.e.
$$x_3 = x_1 - \frac{f(x_1)}{f'(x_1)} \qquad (7.10)$$

Equation (7.10) defines the Newton-Raphson Method. It may alternatively be derived directly by finding the point where the tangent to $y = f(x)$ at the point $(x_1, f(x_1))$ meets the line $y = 0$.

The Newton-Raphson method, being a limiting case of the Secant Method may also fail to converge, but if it *does* converge it does so very quickly as we now prove.

THEOREM. If ρ is the exact solution of $f(x) = 0$ and $x_1 = \rho + \varepsilon_1$ the Newton-Raphson method will converge provided that ε_1 is sufficiently small and

$$\left| \tfrac{1}{2} \varepsilon_1 \frac{f''(\rho)}{f'(\rho)} \right| < 1$$

Proof. Let $x_1 = \rho + \varepsilon_1$ where ρ is the exact solution of $f(x) = 0$. Then (7.9) gives us, as our next estimate of ρ

$$x_2 = x_1 - \frac{f(x_1)}{f'(x_1)}$$

$$= \rho + \varepsilon_1 - \frac{f(\rho + \varepsilon_1)}{f'(\rho + \varepsilon_1)}$$

and so, since $f(\rho) = 0$, if ε_1 is sufficiently small

$$x_2 \doteq \rho + \varepsilon_1 - \frac{\varepsilon_1 f'(\rho) + \tfrac{1}{2}\varepsilon_1^2 f''(\rho)}{f'(\rho) + \varepsilon_1 f''(\rho)}$$

The Solution of Non-Linear Equations

which simplifies, on neglecting ε_1^3 and higher terms

to
$$x_2 = \rho + \tfrac{1}{2}\varepsilon_1^2 \frac{f''(\rho)}{f'(\rho)}$$

Thus
$$|x_2 - \rho| \doteq \left|\tfrac{1}{2}\varepsilon_1^2 \frac{f''(\rho)}{f'(\rho)}\right|$$

and so
$$|x_2 - \rho| < |x_1 - \rho| = |\varepsilon_1|$$

provided
$$\left|\tfrac{1}{2}\varepsilon_1 \frac{f''(\rho)}{f'(\rho)}\right| < 1 \qquad (7.11)$$

This proves the theorem.

If (7.11) is satisfied we will then have

$$|x_2 - \rho| = |\varepsilon_2|, \quad \text{say} = A\varepsilon_1^2$$

so that the error will decrease quadratically. Thus if our first estimate is accurate to 1 d.p., our 2nd should be accurate to 2 d.p., our 3rd to 4 d.p., our 4th to 8 d.p. and so on.

Example 7.3. Use the Newton-Raphson Method to find $\sqrt{2}$ to 4 d.p. starting from $x = 1$.

Solution. In this case $f(x) = x^2 - 2$ so $f'(x) = 2x$. The Newton-Raphson formula is therefore

$$x_{n+1} = x_n - \frac{(x_n^2 - 2)}{2x_n} = \tfrac{1}{2}\left(x_n + \frac{2}{x_n}\right)$$

Hence we have the successive approximations

n	x_n	$|\varepsilon_n|$
1	1	0·4142
2	1·5	0·0858
3	1·4167	0·0025
4	1·4142	0·0000

7.6. The General Iterative Method

The Newton-Raphson method is a particular example of a class of what are known as "iterative methods". An iterative method is

Numerical Analysis

one in which an expression of the form

$$x_{n+1} = F(x_n) \tag{7.12}$$

is used to produce the $(n+1)$st approximation (x_{n+1}) to the solution of the equation

$$x = F(x) \tag{7.13}$$

from the n-th approximation (x_n). The Newton-Raphson method is clearly of this type since it may be written

$$x_{n+1} = x_n - \frac{f(x_n)}{f'(x_n)} \tag{7.14}$$

where $f(x) = 0$ is the equation to be solved.

Under what conditions will a method such as (7.12) converge to a solution of (7.13)? Let us first try a simple example and see what happens.

Example 7.4. Find the positive root of $x^2 - x - 1 = 0$ using iterative methods.

(i) by writing the equation as

$$x = x^2 - 1;$$

(ii) by writing the equation as

$$x = 1 + \frac{1}{x}$$

in each case starting with $x = 1$.

Solution. (i) The iterative form is

$$x_{n+1} = x_n^2 - 1$$

Taking $x_0 = 1$ we have $x_1 = 0$, $x_2 = -1$, $x_3 = 0$, $x_4 = -1, \ldots$ and convergence will not occur.

(ii) The iterative form is

$$x_{n+1} = 1 + \frac{1}{x_n}$$

Taking $x_0 = 1$ we have $x_1 = 2$, $x_2 = \frac{3}{2}$, $x_3 = \frac{5}{3}$, $x_4 = \frac{8}{5}$, $x_5 = \frac{13}{8} = 1 \cdot 625, \ldots$ and convergence to the root $(1 \cdot 618 \ldots = \frac{1}{2}(\sqrt{5} + 1))$ is occurring, although rather slowly.

It is already clear from this example that an iterative procedure may fail to converge. In case (i) the iterated values cycle indefinitely since $x_{n+2} = x_n$ for $n > 1$. If we start with $x_0 = 2$ on the other hand we find that $x_1 = 3$, $x_2 = 8$, $x_3 = 63, \ldots$ and evidently the sequence of iterates is diverging rapidly. The situation is however transformed if we rewrite the equation as in (ii) when convergence occurs.

We obviously need some criterion that we can apply in advance to decide if an iterative method such as (7.12) will converge or not, and this criterion we will obtain in the next section.

7.6. Convergence of the General Iterative Method

THEOREM. If α is the exact solution of (7.13) so that $\alpha = F(\alpha)$, then (7.12) will converge to α from a sufficiently close starting value x_0 if and only if $|F'(\alpha)| < 1$.

Proof. We are using (7.12) so that

$$x_{n+1} = F(x_n)$$

Suppose that $x_n = \alpha + \varepsilon_n$ then

$$\begin{aligned} x_{n+1} &= F(\alpha + \varepsilon_n) \\ &= F(\alpha) + \varepsilon_n F'(\alpha) + 0(\varepsilon_n^2) \\ &\doteqdot \alpha + \varepsilon_n F'(\alpha) \end{aligned}$$

(assuming that ε_n^2 may be ignored, which will be true if x_0 is sufficiently close to α) i.e. $\varepsilon_{n+1} = x_{n+1} - \alpha = \varepsilon_n F'(\alpha)$.

Thus the magnitude of the error at the $(n+1)$st iteration is $|\varepsilon_n F'(\alpha)| < |\varepsilon_n|$ if and only if $|F'(\alpha)| < 1$.

The errors $|\varepsilon_0|, |\varepsilon_1|, |\varepsilon_2|, \ldots$ will therefore form a decreasing sequence if and only if $|F'(\alpha)| < 1$, i.e. convergence will occur if and only if $|F'(\alpha)| < 1$. Q.E.D.

Definition. If an iterative procedure for solving an equation converges to the solution in such a way that the errors ε_n, ε_{n+1} at the n-th and $(n+1)$-st iterations have a relationship of the form

$$\varepsilon_{n+1} = A \varepsilon_n^p$$

then we say that the iterative procedure "*converges with power p*".

Corollary. Since $|\varepsilon_{n+1}| = |\varepsilon_n| |F'(\alpha)|$ the convergence is linear (i.e. $p = 1$) unless $F'(\alpha) = 0$. The smaller the value of $|F'(\alpha)|$ the more rapid the convergence. Thus for fast convergence we should try to arrange (7.12) so that $|F'(\alpha)|$ is small.

Example 7.5. Investigate the convergence of the methods of Example (7.4) in the light of the theorem above.

Solution. $\alpha = 1 \cdot 6180 \ldots$ in this case.

For (i) we have $F'(x) = 2x$ and so $F'(\alpha) = 3 \cdot 2360 \ldots > 1$, so convergence will not occur.

For (ii) $\quad F'(x) = -\dfrac{1}{x^2}$, so $|F'(\alpha)| < \dfrac{1}{2 \cdot 5} = 0 \cdot 4$

and convergence will occur, with the error being reduced by a factor of more than 2 at every iteration.

The theorem shows us how to make the convergence fast, as the next example illustrates.

Example 7.6. Find an integer value, k, so that by writing the equation

$$x^2 - x - 1 = 0$$

in the form

$$x = \frac{1}{k}(1 + (k+1)x - x^2)$$

fast convergence is obtained.

Solution. We have

$$x_{n+1} = \frac{1}{k}(1 + (k+1)x_n - x_n^2)$$

and so, in this case

$$F'(x_n) = \frac{1}{k}(k + 1 - 2x_n)$$

and so $|F'(\alpha)|$ is small if we choose
$$k+1 = 2\alpha = 3 \cdot 32 \ldots$$
so we choose $k=2$ as the nearest integer.

With this choice of k the value of $|F'(\alpha)|$ is about $0 \cdot 16$. If we start with $x_0 = 1$ and use the iteration
$$x_{n+1} = \tfrac{1}{2}(1 + 3x_n - x_n^2)$$
we find
$x_1 = 1 \cdot 5$
$x_2 = 1 \cdot 625$
$x_3 = 1 \cdot 6172$ (to 4 d.p.)
$x_4 = 1 \cdot 6181$ (to 4 d.p.)
$x_5 = 1 \cdot 6180$

and x_5 is correct to 4 d.p.

In general, we cannot know the exact value of $F'(\alpha)$ when we use this method since we do not know the value of α; it is in practice sufficient for us to know an approximate value for α in order to construct a convergent procedure. We illustrate this by solving a transcendental equation.

Example 7.7. Find the solution of $x = e^{-x}$.

Solution. We first observe that there is a solution of the equation in the interval $(0, 1)$.

If we write the iterative equation as
$$x_{n+1} = e^{-x_n}$$
then for convergence we must have
$$|-e^{-\alpha}| < 1$$
where α is the solution; but $\alpha > 0$, so the condition is satisfied. We begin with $x_0 = 1$ and obtain the following results (to 4 d.p.)
$x_1 = 0 \cdot 3679$, $x_2 = 0 \cdot 6922$, $x_3 = 0 \cdot 5005$, $x_4 = 0 \cdot 6062$, $x_5 = 0 \cdot 5454$.

Convergence to the solution $(0 \cdot 5671)$ is occurring, but slowly. The reason for this is obvious, for the errors are reduced at each stage by a factor of $-e^{-\alpha}$, i.e. by
$$-e^{-0 \cdot 5671} \doteqdot -0 \cdot 5671$$
(since $\alpha = e^{-\alpha}$).

Numerical Analysis

Thus our errors will alternate in sign and with an initial error of approximately 0·43 the number of iterations required to achieve 4 d.p. accuracy will be k where

$$(0\cdot 43)(0\cdot 5671)^k = \tfrac{1}{2}\times 10^{-4}$$

the solution of which is $k \doteqdot 16$, and indeed we find, if we continue the iterations from $x_5 = 0\cdot 5454$ then $x_6 = 0\cdot 5796, \ldots x_{10} = 0\cdot 5685$, $x_{15} = 0\cdot 5671$—in good agreement with our prediction.

7.7. Further Comments on the Newton-Raphson Method

From the point of view of speed of convergence the Newton-Raphson method is the most powerful of the 5 methods for solving non-linear equations which we have studied in this chapter. We can summarize the situation as to the power of convergence (as defined in 7.6) as follows:

Method	Power of convergence
Bisection	1·0
False position	1·0
Secant rule	1·618
General iterative	1·0
Newton-Raphson	2·0

The first two methods will always converge to a solution if one exists, the other three may fail to converge, as we have seen. The Newton-Raphson method may fail, or at best converge slowly if the function has a multiple root or two roots very close together. To overcome this difficulty modified versions of the Newton-Raphson method have been developed, one of which we now examine.

Suppose that $f(x)$ has a zero at $x = \alpha$ of multiplicity k. Then $f(\alpha) = f'(\alpha) = \ldots = f^{k-1}(\alpha) = 0$ but $f^k(\alpha) \neq 0$.

We modify the Newton-Raphson formula

$$x_{n+1} = x_n - \frac{f(x_n)}{f'(x_n)}$$

The Solution of Non-Linear Equations

by introducing a parameter, λ, as a factor of the second term, viz.

$$x_{n+1} = x_n - \frac{\lambda f(x_n)}{f'(x_n)} \tag{7.15}$$

and we carry out an analysis to find the best value for λ.

Let $x_n = \alpha + \varepsilon_n$; then

$$f(x_n) = f(\alpha + \varepsilon_n) \doteq f(\alpha) + \varepsilon_n f'(\alpha) + \ldots + \frac{\varepsilon_n^k}{k!} f^k(\alpha) + \frac{\varepsilon_n^{k+1}}{(k+1)!} f^{k+1}(\alpha)$$

and $\quad f'(x_n) = f'(\alpha + \varepsilon_n) \doteq f'(\alpha)$

$$+ \varepsilon_n f''(\alpha) + \ldots + \frac{\varepsilon_n^{k-1} f^k(\alpha)}{(k+1)!} + \frac{\varepsilon_n^k}{k!} f^{k+1}(\alpha)$$

Since α is a zero of multiplicity k these formulae simplify to

$$f(x_n) \doteq \frac{\varepsilon_n^k}{k!} f^k(\alpha) + \frac{\varepsilon_n^{k+1}}{(k+1)!} f^{k+1}(\alpha)$$

$$f'(x_n) \doteq \frac{\varepsilon_n^{k-1}}{(k-1)!} f^k(\alpha) + \frac{\varepsilon_n^k}{k!} f^{k-1}(\alpha)$$

and so, substituting in (7.15) and cancelling

$$x_{n+1} = (\alpha + \varepsilon_n) - \frac{\lambda(\varepsilon_n + a\varepsilon_n^2)}{k(1 + b\varepsilon_n)} \tag{7.16}$$

where $\quad a = \dfrac{f^{k+1}(\alpha)}{(k+1)f^k(\alpha)} \quad$ and $\quad b = \dfrac{f^{k+1}(\alpha)}{kf^k(\alpha)}$

Hence:

$$x_{n+1} - \alpha \doteq \varepsilon_n\left(1 - \frac{\lambda}{k}\right) + \frac{\lambda}{k}(a - b)\varepsilon_n^2 \tag{7.17}$$

We now see, from (7.17) that by choosing $\lambda = k$ we retain quadratic convergence in this modified Newton-Raphson and that

$$\varepsilon_{n+1} = x_{n+1} - \alpha \doteq -\frac{f^{k+1}(\alpha)}{k(k+1)f^k(\alpha)} e_n^2 \tag{7.18}$$

We have therefore proved:

THEOREM. If $f(x)$ has a zero of multiplicity k at $x = \alpha$ the

Numerical Analysis

modified Newton-Raphson formula

$$x_{n+1} = x_n - \frac{kf(x_n)}{f'(x_n)}$$

has quadratic convergence.

Example 7.8. Use the modified Newton-Raphson method to find the double positive root of $x^4 - 4x^2 + 4 = 0$ starting at $x_0 = 1$.

Solution. In this case $k = 2$ so the iterative formula is

$$x_{n+1} = x_n - \frac{2(x_n^4 - 4x_n^2 + 4)}{4x_n^3 - 8x_n}$$

or $\qquad x_{n+1} = \frac{x_n^4 - 4}{2x_n^3 - 4x_n}$. Starting with $x_0 = 1$

we obtain $x_1 = 1.5000$, $x_2 = 1.4167$, $x_3 = 1.4142$ which is correct to 4 d.p.

Had we used the unmodified form of the Newton-Raphson convergence would still have occurred, but much more slowly since (7.17) shows that when $k = 2$ and $\lambda = 1$ convergence is linear and, in fact, $\varepsilon_{n+1} \doteqdot \frac{1}{2}\varepsilon_n$.

Problems on Chapter 7

(1) Attempt to solve the equation $x^3 - 2 = 0$ iteratively, starting at $x_0 = 1$ and using (a) $x = 2/x^2$; (b) $x = (2+x)/(x^2+1)$; (c) $x = (2+10x)/(x^2+10)$. In each case decide whether convergence is occurring or not.

(2) Find an integer value k so that the iterative procedure given by $x_{n+1} = (2 + kx_n)/(x_n^2 + k)$ converges rapidly to $2^{1/3}$.

(3) Show that there is a solution to the equation $x = e^{-x^2}$ lying in the interval $(0.6, 0.7)$. Hence prove that if the equation is written in the iterative form

$$x_{n+1} = \frac{ax_n + e^{-x_n^2}}{1+a}$$

then, for optimal convergence, a should be given a certain value which lies in the interval $(0.7, 1.0)$.

(Cardiff Part 1, 1976)

(4) Show that the polynomial $x^3 - 6x + 3 = 0$ has one root in each of the intervals $(-3, -2)$, $(0, 1)$ and $(2, 3)$. Take $x_1 = \frac{1}{2}$ and use the Newton-Raphson Formula once to find another approximation x_2. Does the root in $(0, 1)$ lie in the interval $(\frac{1}{2}, x_2)$?

(Cardiff Part 1, 1974)

(5) Show that if we attempt to solve the equation $x^x = k$ by using the iterative form $x_{n+1} = (\log k/\log x_n)$ convergence will not occur if k is smaller than a certain number which is approximately equal to 15. Investigate what happens when $k = 16$ if (a) we start with $x_1 = 2$; (b) we start with $x_1 = 3$.

Solutions to Exercises

CHAPTER 2 (1) 0, 1, 2, 3, 4.
(2) 6 d.p.

CHAPTER 3 (1) $311 \cdot 85 \leq \text{sum} \leq 465 \cdot 85$; $0 \cdot 3486 \times 10! \leq \text{product} \leq 2 \cdot 5938 \times 10!$; $\frac{100}{11} \leq \text{sum} \leq \frac{100}{9}$
(2) $1 \cdot 8960 \leq x_1 \leq 2 \cdot 15$, $2 \cdot 80 \leq x_2 \leq 3 \cdot 1540$.
(3) 60, 61·0, 61·66.
(4) 4, 6, 58.
(5) The error in (ii) is more than 10,000 times greater than in (i).

CHAPTER 4 (1) $\frac{1}{12}x(7-x)$; 0·6875; 0·8333.
(2) 0·7071; 0·8660.
(3) $-0 \cdot 0163 x^3 - 0 \cdot 0181 x^2 + 0 \cdot 5344 x$; 0·7059(0·7071).
(4) $x^4 - 5x^2 + 4$; 8·3776.

CHAPTER 5 (1) 4th term should be 1·5166.
(2) 5th term should be $-0 \cdot 125$.
(3) $x^3 + 3x - 8$.

CHAPTER 6 (1) (a) 0·6433, (b) 0·6565, (c) 0·6667.
(2) 1·2189 (correct to 4 d.p.); $|f''''(x)|$ is unbounded over $\langle 0, 1 \rangle$, is <1 over $\langle 1, 2 \rangle$.
(3) 2·0000.

CHAPTER 7 (1) (a) No; (b) Yes; (c) Yes.
(2) $k = 3$.
(3) $x = 0 \cdot 6529$; optimal $a = 2x^2 \doteq 0 \cdot 8525$.
(4) $\frac{11}{21}$; no.

Index

Abel, N. H. 1

Bisection, method of 53–55, 57, 58

Decimal places 5–7
Differences, divided 24, 25
   forward 22–24

Error, absolute 11
   accumulated 10
   relative 12
   rounding 10, 11
   single 33
   truncation 11
   types of 9–11

Error fan 32
Errors, correction of 33–35
   detection of 33–35
   empirical 9
   human 9, 10
   multiple 35, 36

Euler, L. 1
Extrapolation 18

False position, method of 55–59

Gauss, C. F. 1
General iterative method 61–66
   convergence of 63–66

Integration, numerical 39–52
Interpolation 17–29
   linear 18
   by hagrange polynomial 20–22
   by Newton polynomial 25–29
   polynomial 19, 20

Jacobi, C. G. J. 1

Lagrange, J. L. 20
   polynomial 20–22

Midpoint Rule 42, 43
   truncation error 44, 45

Newton, I. 2
   polynomial 25–29

Newton-Raphson method 59–61, 66–68
   convergence of 60, 61

Non-linear equations 53–68
One-point methods 59–68

Pi (π), value of 5

Ramanujan, S. 1
Rounding 6–8
   errors 10, 11

Secant method 57
Simpson's Rule 46–49
   truncation error 44, 45

Trapezium Rule 39–42
   truncation error 44, 45

Truncation error 11
   in Midpoint Rule 44, 45
   in Simpson's Rule 48, 49
   in Trapezium Rule 44, 45

Two-point methods 53–57
   convergence of 57–59